Introduction to Planetary Photometry

This handbook introduces planetary photometry as a quantitative remote sensing tool, and demonstrates how reflected light can be measured and used to investigate our solar system. The author explains how data gathered from telescopes and spacecraft are processed and used to infer properties such as the size, shape, albedo, and composition of celestial objects including planets, moons, asteroids, and comets. Beginning with an overview of the history and background theory of photometry, later chapters delve into the physical principles behind commonly used photometric models and the mechanics of observation, data reduction, and analysis. Real-world examples, problems, and case studies are included, all at an introductory level suitable for new graduate students, planetary scientists, amateur astronomers, and researchers looking for an overview of this field.

MICHAEL K. SHEPARD is Professor of Geosciences at Bloomsburg University of Pennsylvania, specializing in remote sensing, planetary photometry, and asteroid studies. He is the author of the popular science book, *Asteroids: Relics of Ancient Time*, articles for popular science magazines like *Sky and Telescope*, and a guest science columnist for the regional *Press-Enterprise* newspaper. He has previously worked at the Smithsonian Air and Space Museum in Washington, DC and the Jet Propulsion Laboratory in Pasadena, CA.

Introduction to Planetary Photometry

MICHAEL K. SHEPARD
Bloomsburg University

CAMBRIDGE
UNIVERSITY PRESS

University Printing House, Cambridge CB2 8BS, United Kingdom

One Liberty Plaza, 20th Floor, New York, NY 10006, USA

477 Williamstown Road, Port Melbourne, VIC 3207, Australia

4843/24, 2nd Floor, Ansari Road, Daryaganj, Delhi – 110002, India

79 Anson Road, #06-04/06, Singapore 079906

Cambridge University Press is part of the University of Cambridge.

It furthers the University's mission by disseminating knowledge in the pursuit of education, learning, and research at the highest international levels of excellence.

www.cambridge.org
Information on this title: www.cambridge.org/9781107131743
10.1017/9781316443545

© Michael K. Shepard 2017

This publication is in copyright. Subject to statutory exception and to the provisions of relevant collective licensing agreements, no reproduction of any part may take place without the written permission of Cambridge University Press.

First published 2017

Printed in the United Kingdom by TJ International Ltd. Padstow Cornwall

A catalogue record for this publication is available from the British Library.

Library of Congress Cataloging-in-Publication Data
Names: Shepard, Michael K., 1962- author.
Title: Introduction to planetary photometry / Michael K. Shepard (Bloomsburg University).
Description: Cambridge, United Kingdom ; New York, NY : Cambridge University Press, 2017. | Includes bibliographical references and index.
Identifiers: LCCN 2016041181| ISBN 9781107131743 (hardback ; alk. paper) | ISBN 110713174X (hardback ; alk. paper)
Subjects: LCSH: Photometry. | Planets–Observations.
Classification: LCC QC391 .S54 2017 | DDC 522/.62–dc23 LC record available at https://lccn.loc.gov/2016041181

ISBN 978-1-107-13174-3 Hardback

Additional resources for this publication at www.cambridge.org/photometry.

Cambridge University Press has no responsibility for the persistence or accuracy of URLs for external or third-party Internet Web sites referred to in this publication and does not guarantee that any content on such Web sites is, or will remain, accurate or appropriate.

For Samantha, Ellen, Alexander, and Benjamin

Contents

		Preface	*page* ix
		Acknowledgments	xi
		A Note on the Symbols and the Text	xii
1		**A Brief History of Planetary Photometry**	1
	1.1	Photometry Defined	1
	1.2	The Nature of Vision and Light	2
	1.3	The Human Eye and Visual Magnitude	5
	1.4	Early Principles of Photometry	10
	1.5	Early Photometers	14
	1.6	The Brightness and Color of Stars and Planets	20
	1.7	What Was Old Is New Again	23
2		**Photometry Conventions, Terminology, and Standards**	30
	2.1	Coordinate Systems	30
	2.2	Time	39
	2.3	Photometric Nomenclature	45
	2.4	Magnitudes	50
	2.5	Photometric Systems	54
	2.6	Standard Stars, Solar Analogs, and the Zero Point Magnitude	60
3		**The Mechanics of Planetary Observing**	64
	3.1	Telescopes	64
	3.2	Optical System Properties	70
	3.3	Older Detection Systems	74
	3.4	Charge-Coupled Devices	76
	3.5	Measurement and Reduction	87

Contents

4	**The Physical Basis of Photometric Scattering Models**	95
	4.1 The Nature of Light	95
	4.2 Refraction, Fresnel Reflection, and Absorption	99
	4.3 Scattering by Particles	105
	4.4 Particle Phase Function	107
	4.5 Scattering by Ensembles of Particles and Radiative Transfer	113
	4.6 Opposition Surge	121
	4.7 Surface Roughness	126
5	**Planetary Reflectance and Basic Scattering Laws**	131
	5.1 Lambertian Reflectance	132
	5.2 Disk-Resolved Albedos and Measures of Reflectance	135
	5.3 Reciprocity	140
	5.4 Disk-Resolved Scattering Models and Behavior	144
	5.5 Disk-Integrated Albedos and Measures of Reflectance	150
	5.6 Disk-Integrated Scattering Models and Behavior	155
	5.7 More Complex Photometric Models	160
	5.8 Extracting Parameters from Complex Models	167
6	**Planetary Disk-Integrated Photometry**	170
	6.1 Planetary Size and Albedo	170
	6.2 Lightcurves	174
	6.3 Phase Curves of Planets	180
	6.4 Phase Curves of Small Bodies	185
	6.5 Polarization Phase Curves	195
	6.6 Case Study: Mercury	200
7	**Planetary Disk-Resolved Photometry**	204
	7.1 Laboratory Methods and Tests	204
	7.2 Spacecraft Observations	208
	7.3 Case Study: Laboratory Tests of the Hapke Model	213
	7.4 Case Study: Mercury and MESSENGER	217
	7.5 Case Study: The Moon and Clementine	224
	Appendix 1: Problems	229
	References	234
	Index	247

Preface

Planetary photometry is the science that deals with measuring the brightness of an object within the solar system, determining how that brightness varies with illumination and viewing geometry, and discovering what this can tell us about the object. With the exception of the rare physical specimens we obtain, we see planets only through the eyes of telescopes or the cameras of our robotic emissaries. It is an old and revered field, but it is arcane. Much of the field is based upon papers written more than a century ago, with methods, units, and terminology that were originally based upon the needs of stellar astronomy, and only later applied to lunar and planetary studies. There is a confusing array of terms that mean different things to different people. Like old cities, the design is ad hoc, and if we could do it over again, we would probably design things differently. But it is what it is, and any who want to use or understand it must undergo an initiation into its argot.

There are few books that deal with the subject for the practicing planetary scientist. Hapke's (1993, 2012) classic is the best reference, but it would rightly be considered a somewhat advanced text, and there are topics of practical interest that it doesn't include. As a graduate student and later as a practicing scientist in the field, I found that much of what I wanted to know could only be found by reading dozens, if not hundreds of papers. Like Samuel Johnson, I found that "A man will turn over half a library to make a book."

Planetary photometry is not a sexy science, but it undergirds nearly every planetary discovery based on the use and comparison of telescopic and spacecraft observations. And though it seems innocuous, more than one researcher has been bitten by not considering how lighting and viewing might affect their observations. I wrote this book to introduce the basic concepts and principles and serve as a guide, if needed or desired, for more advanced studies.

I assume only modest mathematical background, equivalent to the first year or so of college mathematics. I discuss the physics behind many of the concepts of planetary photometry, but generally do not delve deeply into them. Other books do that well already. However, because the derivations for many important photometric equations are rarely included in the literature, I walk the reader through many of these – a type of hand-holding that I often wished for when first encountering some of this material. I illustrate many concepts with examples and include a number of problems, with answers available online at www.cambridge.org/photometry.

After digesting this book, you should be able to converse with experts in planetary photometry without embarrassment, and have a good sense of the potential limitations of your or others data where photometry plays a role. And if the desire should take, you should be ready to tackle more advanced books on the subject.

Acknowledgments

I am the beneficiary of many fruitful discussions with a long line of experts in this field, chief among them Paul Helfenstein and Bruce Hapke. And although I have never worked directly with them, I have greatly benefited from reading the works of Yuri Shkuratov and his many colleagues at Kharkov University, and am grateful for their long tradition of excellent photometric and polarimetric work. Similarly, work by Kari Lumme, Ted Bowell, Jay Goguen, Joe Veverka, Kari Muinonen, Jian-Yang Li, Michael Mishchenko, and Bonnie Buratti are nourishing staples in my photometric library. I would also like to thank Ray Arvidson, Bruce Campbell, Ed Guinness, Jeffrey Johnson, Will Grundy, Deborah Domingue, David Paige, Emily Foote, Bob Nelson, Patrick Pinet, and Ed Cloutis as colleagues in a variety of my work in this field over the years. I am also indebted to the marvelous librarians at Bloomsburg University: Andrea Schwartz, Charlotte Droll, Linda Neyer, and Michael Coffta.

I am grateful to a number of people who took time to read over portions of this book: Alan Harris, Brian Warner, Jon Giorgini, Joe Masiero, Kjartan Kinch, Jeff Johnson, Jian-Yang Li, Bruce Hapke, and Vishnu Reddy. Despite their attention, any mistakes remaining are my own.

Many thanks to The Royal Observatory at Edinburgh and Steve Larson at the Catalina Sky Observatory for allowing me to use their images.

A Note on the Symbols and the Text

The mathematical symbols used by the host of referenced works are often inconsistent between writers. While some symbols have been used repeatedly and become traditional, others have not. I have attempted to be self-consistent, but note that there are simply too many physical variables for the limited number of standard Roman and Greek letters available. In several instances, I have reused symbols. In those cases, I have attempted to make the meaning clear to prevent confusion.

Throughout the text, a number of words are in **bold**. This is to indicate a term of some importance. Proper names are also in **bold** the first time they are introduced. *Italics* are used to emphasize a point; they are also used for the title of books or papers being discussed.

1
A Brief History of Planetary Photometry

If you've picked up this book, odds are that you are a scientist, engineer, amateur astronomer, or student who is interested in the study of planets using images acquired either with a telescope or spacecraft. Even in the modern era, the vast majority of what we know about objects in the solar system comes not from physical samples but from images. For centuries this was all we had, and our predecessors labored to learn as much as possible from subtle differences in brightness and color. As we noted in the preface, the sheer antiquity of the subject has left many parts of the field arcane, and so throughout this book we will, where possible, introduce concepts in historical context. In this chapter, we briefly introduce the historical backdrop and development of the field up to the beginning of the twentieth century, and we end with a recent example of its power and continuing applicability. But first, we must make it clear what it is we are talking about.

1.1 Photometry Defined

This book is about planetary photometry, but technically this is a misnomer. **Photometry** is defined as the science of the measurement of light in terms of how the human eye perceives it. But because human eyes differ in their sensitivity to colors and the intensity of perceived brightness, this is a subjective measurement that must be standardized. True photometry is used by lighting engineers to determine how best to light a room, street, or building. This is not what astronomers or planetary scientists mean by photometry.

The real subject of this book is **radiometry**, the measurement of light in standard physical units, independent of the perception of the human eye. However, early astronomy, including planetary science, was strictly a visual science and all of the early measurements of stellar and planetary brightness and color relied upon the human eye – it *was* photometry. Today, astronomers still use and publish *visual* magnitudes, or those determined after using a filter and light detector that approximates the wavelength range of the eye, but they have removed the physiological basis of measurement.

Despite these changes, the term photometry has persisted in planetary science. With *telescopic* observations, the term **photometric calibration** is used to mean the reduction of data using standard stars and color filter sets, such as the Johnson-Morgan UBV system (see Chapter 2). *Spacecraft* missions will often use the term **radiometric calibration** to mean that their images have been calibrated into standardized physical units, e.g. W m^{-2} sr^{-1} μm^{-1}. However, if a scattering model is used to correct those calibrated values to what would be observed under a standardized lighting and viewing geometry (e.g. looking straight down on a planet, the Sun 30° from zenith), the results are also sometimes said to have been photometrically calibrated.

To summarize, in common usage **planetary photometry** means the measurement and calibration of light from distant non-stellar objects using some system of standard references, and where possible, correction (or normalization) to common illumination and viewing circumstances.

1.2 The Nature of Vision and Light

1.2.1 Early Theories of Vision

Early theories of light and vision generally postulated one of two main ideas: light or an equivalent originated in the eyes and traveled outward where it interacted with objects, the **extromission theory**; and the opposing view that light or an equivalent left the objects and entered the eye, the **intromission theory**.

One of the earliest treatises on light and optics, called *Optics and Catoptrics*, is that of Euclid (circa 300 BCE), who is most famous for his book *The Elements* (of geometry). Like *The Elements*, *Optics* begins with axioms on the behavior of light and derives more complex ideas. Euclid subscribed to the extromission concept of vision (DiLaura, 2006). Like much of early Greek science, this work contained little or no experimentation to test or check concepts.

Ptolemy of Alexandria (90–168 CE), best known for his *Almagest* and the Earth-centered model of the solar system, wrote a treatise some four centuries after Euclid called *A Work on Optics*; it similarly assumed that vision originated in the eye. Unlike Euclid, however, Ptolemy's work included early experiments with refraction (DiLaura, 2006).

Perhaps the most influential book on optics in the Middle Ages, if sometimes indirectly, was the *Book of Optics* (Kitab al-Mazir) by the great Arabic scientist **Ibn al-Haytham** (aka **Alhazen**) (965–1040 CE). Alhazen subscribed to the intromission concept of vision, correctly arguing that, if the eyes were the source of some type of illuminating rays, we could see in the dark. He correctly separated the concept of light from vision, and experimentally demonstrated that light travels in straight lines, and that different sources of light do not interfere with each other, a phenomenon referred to as *immiscibility* (DiLaura, 2006; Falco and Weintz Allen, 2008).

Alhazen's work was inaccessible to many until the Archbishop of Canterbury and scholar **John Peckham** (1230–1292) wrote *Perspectiva communis* circa 1260. Lindberg (1981) notes that this book was not a presentation of new ideas, but primarily an orderly restatement of much of Alhazen's work, and where possible, other work in the field.

By the time of **Johannes Kepler** (1571–1630), the intromission concept was firmly entrenched, the eye was recognized as an optical device, and the mathematics of geometry was employed to study reflection and refraction. In his 1604 treatise *Ad Vitellionem Paralipomena Quibus Astronomiae pars Optica Traditur* (Supplement to Vitello on the Optical Part of Astronomy), Kepler advanced the modern conception of light rays and recognized, if not explicitly, the inverse square law for the attenuation of light over distance (Lindberg, 1981; DiLaura, 2006; Malet, 2010). The law would not be stated explicitly until 1634 by the French theologian and mathematician **Merin Mersenne** (1588–1648) (DiLaura, 2006).

1.2.2 Modern Concepts of Light

By the seventeenth century, there were two competing theories on the nature of light. In one corner, **Christiaan Huygens** (1629–1695), a Dutch physicist and astronomer, proposed in his 1690 *Treatise on Light* that light was a wave phenomenon like sound. As light traveled, it excited a new spherical wave at each point in its path; these waves add to give the wave front phenomenon we see (Huygens, 1900 [1690]). Today, these excited waves are often referred to as *Huygens wavelets*. This theory differed dramatically from the previous conception of light as a ray-like thing that traveled in straight lines, but was later found to be consistent with experimental evidence.

The competing theory, championed by **Isaac Newton** (1642–1726) and published in his 1704 *Opticks*, was that light was of a *corpuscular* nature – composed of particles that traveled in straight lines. Unlike Huygens's conception, this *was* consistent with the ray theory of optics. These corpuscles were thought to be objects of tiny size with a pure color, and Newton used this concept to explain why white light split into different colors as it traversed a prism (Newton, 1704; DiLaura, 2006).

By the late seventeenth century, it was known that the speed of light was finite, if rapid. The first indications of this came from observations of Jupiter's moon Io by the Danish astronomer **Ole Roemer** (1644–1710) in 1676 (Roemer, 1677). Roemer's method is quite ingenious. The inner moon Io orbits Jupiter every 42.46 h. If the distance between Earth and Jupiter were constant, an observer would see Io disappear behind Jupiter or reappear from the other side every 42.46 h. However, depending upon our relative orbital positions, the Earth may be approaching or receding from Jupiter. If approaching, the apparent orbital period appears to be less than 42.46 h; if receding from Jupiter, the apparent period is slightly longer. Roemer realized that the differences in apparent period were caused by the change in distance between Earth and Jupiter, and that light took a finite amount of time to traverse that change in distance (Mach, 1926). Although it is often reported that Roemer estimated the speed of light from this work, French (1990) states that, for whatever reason, Roemer did not, but Huygens did in 1690.

By 1800, other optical phenomena were garnering attention. A number of scientists had noted that the shadows cast by edges were not sharp, but consisted of alternating bright and dark bands. Similar bands were seen in the shadows cast when light passed through double slits. In 1804, **Thomas Young** (1773–1829) performed a series of classic experiments that demonstrated that these features were caused by an **interference effect**, analogous to the interference of ripples in water, and best explained if light were wavelike in nature (Young, 1804). Despite Newton's status in the worlds of physics and astronomy, the wave theory championed by Huygens became the standard model of light until the twentieth century because only it could easily explain these observations. Despite this, the ray theory still enjoyed (and currently enjoys) popularity, especially for modeling the behavior of light in systems of mirrors and lenses. The modern conception of light behaving as both wave and particle would not arise until the early twentieth century and quantum mechanics.

In 1819, the French physicist **Augustin-Jean Fresnel** (1788–1827) published a *Memoir on the Diffraction of Light* in which he adopted Huygens principle of wavelets to explain the experiments of Young, a phenomenon now referred to as **diffraction** (Fresnel, 1819). This marriage of ideas was extremely powerful and is still a useful way to think about diffraction. Fresnel later teamed up with **François Arago** (1786–1853), another French physicist, to show that polarized light did not always behave as expected in these diffraction experiments (Arago and Fresnel, 1819).

Polarized light had been discovered more than a century earlier when **Rasmus Bartholin** (1625–1698) passed a narrow beam of light through a crystal of Iceland spar, a form of calcite (Bartholin, 1670). After passing through the calcite, the point was split into two beams, a phenomenon referred to as **double refraction**. Materials with this property are called **birefringent**, and today we recognize it to be a consequence of an anisotropic refractive index. The two beams, often called the *ordinary* and *extraordinary rays*, are polarized in perpendicular directions. What Arago and Fresnel demonstrated was that, while rays of the same polarization interfered to produce fringes, perpendicularly polarized rays did not.

1.3 The Human Eye and Visual Magnitude

Early photometry relied exclusively on the only sensor available to measure brightness – the human eye, a remarkable organ sensitive to a narrow range of the electromagnetic spectrum. To understand why photometry developed the way it did, we need an understanding of how the eye works.

1.3.1 The Sun's Light and Planck's Law

The eye has evolved with its particular spectral sensitivity characteristics because of the circumstances of our place in the universe. We orbit an average star, classified as a G2-type (yellow, main-sequence) star, with a "surface" temperature of about 5,800 K.

All objects generate and radiate electromagnetic energy; how much and at what wavelength is principally determined by how hot they are. The physicist **Max Planck** (1858–1947) derived a formula, now known as the Planck Function, that allows us to predict how much energy is radiated from a **blackbody** (a perfectly efficient radiator or *emitter*) at any given temperature and wavelength of light:

$$L(\lambda, T) = \frac{2hc^2}{\lambda^5} \left[\exp\left(\frac{hc}{\lambda kT}\right) - 1 \right]^{-1} \qquad 1.1$$

Here L is the emitted spectral radiance[1] (power per unit area per wavelength interval in the normal direction per unit solid angle), h is Planck's constant

---

[1] We will define this quantity in Chapter 2. For now it is sufficient to note that it can be equated to "brightness."

6   A Brief History of Planetary Photometry

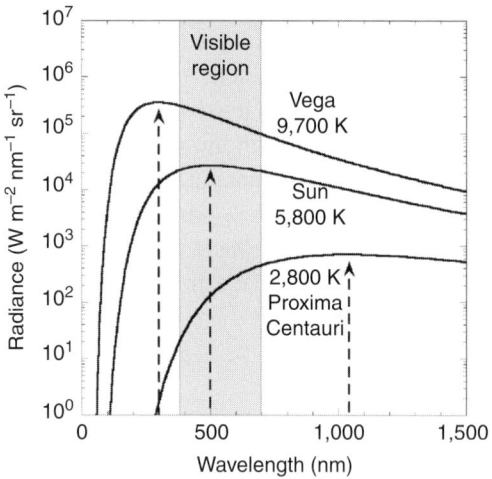

Figure 1.1. Blackbody curves for surfaces with temperatures equivalent to those of the Sun, Vega, and Proxima Centauri.
Credit: Michael K. Shepard.

(6.626×10^{-34} J s^{-1}), c is the speed of light (3×10^8 m s^{-1}), λ is the wavelength (m), k is the Boltzmann constant (1.381×10^{-23} J K^{-1}), and T is the temperature of the object (K).

Curves of this function for a given temperature are called **blackbody curves**. In Figure 1.1, note that for an object at the temperature of the Sun, the peak of the emitted radiation is at a wavelength of 500 nm, right in the part of the electromagnetic spectrum that our eyes see as yellow-green. Not coincidentally, this broad region is where our eyes are most sensitive to light. It evolved that way to take maximum advantage of the ambient solar lighting on Earth. If we had evolved around a star with a surface temperature of 9,700 K, such as Vega, our eyes would likely be most sensitive around 300 nm, deep within the ultraviolet region. Or if creatures with sight have evolved around the red dwarf Proxima Centauri, our nearest stellar neighbor, they are likely to be most sensitive to wavelengths around 1,040 nm, well within the near-infrared.

1.3.2 Physiology and Perception

The eye is sensitive to an incredible dynamic range of luminance, from high noon on Arctic ice to the midnight darkness of the wilderness, illuminated – if at all – only by starlight. There is more than a factor of a billion in the

1.3 The Human Eye and Visual Magnitude

illumination between these environments (Kitchin, 2003). To deal with a range this large, the eye uses several mechanisms.

The pupil of the eye is a variable aperture that closes and opens in response to the amount of light available. Its aperture ranges from ~1 mm in daylight to about 7 mm when fully dark-adapted. Thus the possible change in total light incident on the retina is a factor of $~7^2 = 49$; or alternatively, the daytime eye only lets in about 2% of the nighttime eye. If the only restriction on the light input to an electronic sensor was an aperture that could reduce the amount of incident light by 98%, those capable of measuring variations at night would tend to saturate or fail outright if used in the daytime. To prevent this and still remain sensitive to a wide range of incident light, the eye uses additional mechanisms.

First, we have two different light sensors, **rods** and **cones**. Cones require abundant light to activate and essentially work only in daylight or other well-lit areas. They are tightly packed in a ~1 mm wide region, called the **fovea**, in the center of the **retina**, the web of light-sensitive material that covers the back of the eye. Cones come in three different versions sensitive to different frequencies of light – roughly, blue (short wavelength or S-cones, peak sensitivity at ~430 nm), green (medium wavelength or M-cones, peak sensitivity at ~540 nm), and red (long wavelength or L-cones, peak sensitivity at 580 nm). Thus **photopic**, or cone-based, vision is the high-resolution color vision we experience when looking *directly* at something. Overall, the photopic eye is most sensitive to light around 550 nm because the M- and L-cones together dominate over the S-cones (Kitchin, 2003).

As light levels drop, cones lose their effectiveness, and rods, receptors about one hundred times more sensitive, take over (Kitchin, 2003). Vision based only on the rods is called **scotopic**. Rods are sparse in the fovea, but cover the rest of the retina. They are a broad-band, monochromatic sensor, so in low light, we see only in shades of gray. They are most sensitive to light in the blue-green region of the spectrum (peak around 500 nm), so at moderately low light levels, like dawn or dusk, there is a shift in perceived colors toward the blue as we use both rods and cones. This is called the **Purkinje effect**, after Czech scientist **Jan Purkyne** (1787–1869), and is known to affect visual estimates of stellar magnitudes when comparing stars of different colors (assuming the telescope intensifies the starlight enough to keep cones partially active; Rossotti, 1983).

In very low light, we also use a biological pigment called **rhodopsin**, or visual purple because of its purple tint. In bright light conditions, it is photobleached, or destroyed by light. In the absence of light, it regenerates, taking about 30–45 minutes to fully form and for the eye to become dark-adapted.

When a photon of light hits a rod coated in rhodopsin, it sets off a chemical cascade, triggering the nerve and greatly increasing the sensitivity of the rods to light. It is this dark adaptation that allows humans to see reasonably well in low-light conditions. However, because the dark adaptation is rod-based, the fovea is insensitive under these conditions. That is why those trying to detect faint objects will look slightly to the side of the suspected object to move it off the fovea and onto the rod-covered part of the retina.

For photometry, an important consideration is the contrast sensitivity of the eye. Given two adjacent sources of illumination, how different must they be for the observer to detect them as separate? In the mid-nineteenth century this type of work became the foundation of a branch of experimental psychology called psychophysics – the relationship between physical stimuli and their perception, or more loosely, the mind-body connection. It began when the German physician and experimental psychologist, **Ernst H. Weber** (1795–1878) conducted experiments that led to the empirical relationship now known as **Weber's law**. Weber found that when people are subjected to two sensory stimuli, there is a value called the **just-noticeable difference** that is required before the subject can distinguish between the two stimuli. For example, suppose a subject is holding masses of 100 g in each hand, and the mass in one hand is slowly increased. At some point, let us say 102 g, the subject will notice that the two weights are now different. The just-noticeable difference is 2 g. But if the weights are 1,000 g, a difference of 2 g will not be noticed. Weber found that one needed to add 20 g in that case for the difference to be noticed. Weber's law states that the *ratio* of the just-noticeable difference to the stimuli is a constant; in this case, a 2% difference in weight must be applied for it to be noticed (Weber, 1834).

This behavior was rediscovered and quantified by the German physicist and experimental psychologist **Gustav T. Fechner** (1801–1887) who recast it into a mathematical relationship called **Fechner's law** or the **Weber-Fechner law**. Over some range of magnitudes, there is often a logarithmic relationship between the physical intensity of a stimuli and its sensation

$$S = k \log I \qquad 1.2$$

where S is the perception of the stimuli in some arbitrary units, k is a constant that must be experimentally determined, and I is the physical intensity of the stimulus (Masin et al., 2009). As with anything involving perception, this "law" is only an approximation. It does, however, nicely coincide with our perception of brightness and the stellar magnitude scale.

Well before Fechner, Bouguer (1729) experimentally determined that the eye could just distinguish between light stimuli that differed by 1 part in 64, or

1.6% (Mach, 1926). Subsequent work finds that over a wide range of illumination intensities, a value of 1.8% is often more appropriate (Walsh, 1926).

1.3.3 Early Magnitude Systems

Early astronomers, beginning with Ptolemy, classified the stars in the sky according to their apparent brightness. The brightest stars were classified as being of the first **magnitude**. Here "first" refers to the highest or greatest. Slightly dimmer stars, roughly half as bright, were classified as second magnitude, and so on. The dimmest stars that could be perceived were classified as sixth magnitude. It is an unfortunate system because brighter stars have smaller magnitudes. There have been modern attempts to change the system, but the inertia of history has firmly entrenched it (Hearnshaw, 1996).

Originally, this classification scheme had little or no quantitative underpinning. This made stellar magnitudes a matter of judgement, and there was no way to include either the brighter objects, including the visible planets, Moon, or Sun, or fainter objects once the telescope was invented. These problems were finally tackled by astronomers in the late eighteenth and nineteenth centuries as they attempted to cast the system into a more quantitative framework. There were several issues that required attention.

The first major problem to be confronted was that the eye is not a linear detection system. The English astronomer **John Flamsteed** (1646–1719) noted that stars that differed by a magnitude by eye also differed by a magnitude even when made brighter by viewing through a telescope. The Swedish physicist and astronomer **Anders Celsius** (1701–1744) invented an early extinction photometer and found that stars that differed by a magnitude when viewed in a telescope also differed by a magnitude even when an absorbing plate was inserted into the optical train (Hearnshaw, 1996). Both observations reveal that the eye is not responding linearly, but geometrically to brightness – that it is the *ratio* of brightness differences that are detected, and stars that differ by a magnitude have some multiplicative difference in brightness, not an absolute difference. As we noted earlier, this was also experimentally demonstrated by Weber and Fechner in the mid-nineteenth century. The question was, what was the multiplicative factor? Hearnshaw (1996, in his table 3.1) lists more than a dozen values experimentally determined from 1829 to 1888; these factors range from a low of 2.241 by Fechner to a high of 2.8606 by **P. Ludwig Seidel** (1821–1896). This uncertainty in the true factor made it difficult to standardize stellar catalogs.

1.3.4 The Pogson Interval

In 1856, the English astronomer **Norman Pogson** (1829–1891) proposed that the factor be fixed by fiat, so that two stars with a brightness difference of one magnitude have a brightness ratio of 2.512, or

$$\frac{E_A}{E_B} = 2.512 \quad \text{if} \quad m_B - m_A = 1.0 \text{ mag} \qquad 1.3$$

Here, E is the brightness, or irradiance of incident starlight (the incident light is effectively collimated), and m is the apparent magnitude of the star. There were advantages of this odd ratio. It was a close average to the ratio measured by previous astronomers, and it was mathematically convenient; if two stars have a brightness ratio of 100, they differ by 5 magnitudes ($2.512^5 = 100$) so that a sixth magnitude star is 100 times fainter than a first magnitude star. Hearnshaw (1996) notes that none other than **Edmund Halley** (1656–1742) had presciently stated just this relationship in 1720.

By the late nineteenth century, Pogson's proposal had been adopted by most of the world's major observatories (Hearnshaw, 1996). It is now thought that the eye's response to brightness is better described by a power-law than a logarithm, but Pogson's proposal is now as deeply embedded in the definition of magnitude as is the reversed scale.

1.4 Early Principles of Photometry

Photometry developed along the lines that it did because the eye was the only instrument available to detect and qualify concepts like bright and dark, and as we have seen, there are caveats in its use as a quantitative instrument. In this section, we review the major contributors and concepts that enabled its development.

1.4.1 Concepts and Contributors

Perhaps the most important concepts for the development of modern photometry are those of the light ray and ray density. The ray is a helpful fiction, a way to describe the idea that something with energy travels linearly from point to point. The fact that light traveled only in straight lines had been known at least since Alhazen, who demonstrated this experimentally. However, there was widespread confusion over why light got weaker with distance. Throughout the Middle Ages, scholars thought that the rays themselves weakened with distance (DiLaura, 2001). However, a much better way of thinking about the diminution of light was introduced by **Francesco Maurolico** (1494–1575), a Benedictine monk and professor of mathematics in Messina, Italy. In addition to being one of the first to accurately describe the causes of myopia (near-sightedness) and presbyopia (far-sightedness) (Ilardi, 2007), he is also one of the first to consider light intensity to be a function of the *density* of rays (DiLaura, 2006). In this conception the total number of rays – a proxy for the total quantity of light – is conserved. Additionally, the rays are all the same and do not lose energy with distance unless they are absorbed in a medium. Instead, many rays emanate from a point source (on a surface, there are many point sources), and the *density* of those rays decrease as the ray bundle expands with propagation distance. Brighter sources start with more rays than dim ones. As it turns out, this concept is key to understanding many of the principles of photometry (DiLaura, 2001).

In 1729, the French physicist and mathematician **Pierre Bouguer** (1698–1758), perhaps best known for his work in geodesy and geophysics, published *Essai d'optique sur la gradation de la lumiere* (*Optical tests on the dimming of light*), a work devoted to the measurement of light. In this work, Bouguer stated or foreshadowed many of the concepts and developments that would result in modern photometry. Among the accomplishments of this work, Bouguer experimentally determined that the Sun is 300,000 times brighter than the Moon – an excellent first estimate given that actual difference is 14 magnitudes, or a factor of 400,000.

Although Bouguer laid much of the initial groundwork, the true watershed moment in the history of photometry was the publication of *Photometria* (in Latin) by the German/Swiss scientist **Johann Heinrich Lambert** (1728–1777). The title of this book, coined by Lambert, is the source of our term photometry – literally photo (visual light) – metria (measurement). DiLaura (2001, 2006) notes that Lambert was less interested in theory and deep understanding than in *phenomenology* – observing the behavior of light and deriving mathematical relationships that could be used to describe and quantify these observations.

By now, it was recognized that there was a difference between the physical stimulus of light, or **luminance**, and the perception of brightness. A candle seen at night would be perceived as bright; however, the same candle at the same distance observed in the daytime might not be perceived at all. As a result, the eye is not a reliable instrument for quantifying the brightness of a solo source, or even the difference in brightness between two different sources. However, if two sources are seen under the same circumstances, the eye is very good at detecting any difference between them. This was the key to developing a quantitative system of photometry – using the eye as a **null instrument**. In a null instrument, conditions are adjusted until two different stimuli are indistinguishable. The relative difference of the two sources is then derived from a knowledge of the conditions needed to get them to that state. We will look at this more explicitly in the next section.

1.4.2 Laws of Photometry

Lambert's *Photometria* is considered a defining publication in photometry for a number of reasons. It was based on experiment and observation; the work describes 40 experiments, most of which Lambert conducted himself in the course of his investigations. Lambert systematized, more than any before, much of the vocabulary of photometry into one self-consistent work. And the work was heavily quantitative, using mathematics far more than similar works of that period, both to calculate and estimate error (DiLaura, 2001). The interested reader is strongly encouraged to consult the excellent translation and introductory notes of the *Photometria* (Lambert, 1760) by DiLaura (2001). Here we summarize the four photometric laws discussed by Lambert, and briefly mention some of the other notable points to take from his book.

1.4.2.1 First Law – Immiscibility and Linear Addition

The first widely accepted principle of photometry is that light does not mix, and as a result, sources add their intensity linearly. This principle, called *immiscibility*, was introduced and experimentally demonstrated by Alhazen. Two candles light a wall twice as brightly as a single candle. As Young showed later, this does not hold when the same ray is split and recombined with itself (interference), but for the photometric applications considered by all the authors we will consider, unrelated rays of light sum linearly and do not interfere.

1.4.2.2 Second Law – Inverse Square Law

A second principle of photometry is that the illuminance (total incident power per unit area) falls off with the square of its distance from a source, also known as the inverse square law. As noted earlier, this concept was known, though not directly stated, as far back as Kepler. The first explicit statement of this principle is attributed to Marin Mersenne (DiLaura, 2006). In his posthumously published work *L'Optique et la Catoptrique* (1651) he also explains this principle in terms of light rays emanating from a point source; as a cone of light moves away from the source, the number or rays within the cone is fixed, but their density decreases with the square of the distance because the surface area of an enveloping sphere increases with the radius square.

1.4.2.3 Third Law – Cosine Incidence Law

Lambert, following Maurolico, championed the concept of light density, or *luminous intensity*, which follows from the idea that light is contained in rays, and the number of rays per unit area are a measure of intensity. This concept leads us directly to the third law: the illuminance (power per unit area) is proportional to the cosine of the incidence angle (Lambert's original law stated the sine of the incidence angle, but his angle was the complement of that in use today). If the incidence angle is non-normal, the rays are spread out over a larger surface area that is proportional to the cosine of the angle. This law is referred to today as **Lambert's cosine incidence law**.

These three laws are the fundamentals of Lambert's system of photometry. Mach (1926) notes that if any of these three are assumed as axioms, the other two can be verified experimentally – they are internally self-consistent, and in some sense, different ways of stating the same principle. He also notes that all three of these laws were previously known by others, if implicitly, including Bouguer.

1.4.2.4 Fourth Law – Cosine Emission Law

Lambert also included a fourth photometric law, which states that the intensity of light observed from a diffusely reflecting or emitting surface is a function of the cosine of the emission or viewing angle, a phenomenon referred to as **Lambert's cosine emission law**. This law, while convenient for many calculations, is not generally true for all surfaces. Those surfaces that *do* obey this rule are called **Lambertian reflectors** or **Lambertian emitters** and will be discussed at great length in Chapter 3.

1.4.3 Additional Concepts and Problems

Lambert also examined the role of attenuation due to absorption in translucent materials, and credits Bouguer with finding the result that the loss of brightness is in proportion to the path length through the material (Lambert, 1760; DiLaura, 2001). For example, if the brightness of a light source is reduced by a factor of two when traversing a certain thickness of an absorbing media, two thicknesses will reduce the brightness by a factor of four, etc. This geometric progression gives rise to an exponential decay function. A related version of this behavior is attributed to the German chemist **August Beer** (1825–1863) who noted that the absorption of light through a solution was proportional to the concentration of solute and the path length. Although Bouguer was the first to discuss this behavior, the law is usually referred to as **Beer's law** or the **Beer-Lambert law**.

The main photometric problem not solved by Lambert is that of color. He recognized that the eye is more sensitive to some colors than others so that perceptions were not always proportional to actual intensity; a red and blue source could not be nulled. He did, however, describe experiments that foreshadowed techniques of spectrophotometry (DiLaura, 2001), but the practical development of this field would wait for the twentieth century. In the meantime, there were ad hoc fixes that included using color filters to make the colors of sources approximately the same for easier comparison.

1.5 Early Photometers

1.5.1 The Impetus to Develop Photometric Instrumentation

Early photometry developed along two sometimes overlapping branches during the late eighteenth to twentieth centuries: scientific photometry and industrial photometry (DiLaura, 2006). The industrial branch was driven by the need to quantify the cost of a unit of lighting brightness. As coal-gas and electrical lights began to displace older light sources like candles, it became a matter of economics to determine the least expensive way to illuminate buildings and streets. The scientific branch was driven by astronomy and the desire to quantify the brightness of stars, the Sun, Moon, and planets. It was hoped that this would allow scientists to estimate quantities like the distances of stars or the intrinsic reflectivity of planets – their **albedo**. Both branches were driven to develop measures to quantify brightness and the instruments to make these measurements.

One of the difficulties encountered in these years was the lack of a good way to measure light intensity. During this same period, physicists had developed thermometers to measure heat, but efforts to develop a photometer fell short. The only instrument capable of measuring light was the human eye, and it was notoriously fickle and unreliable in any absolute sense. It was more reliable if used in a relative sense by comparing the apparent brightness of two objects, but it was at its best if conditions could be arranged so that the apparent brightness of two objects could be made equal (i.e. null their difference), and their relative brightness then deduced from the arrangements necessary for this to occur. It was in this latter capacity that concepts in Lambert's *Photometria* were invaluable.

1.5.2 Basic Operation of a Photometer

Two different designs for practical photometers were developed in the eighteenth and nineteenth centuries. An **extinction photometer** uses absorbing plates or other methods to reduce the brightness of an object until it can no longer be detected; the characteristics of the reduction are used to estimate the brightness. For example, the astronomer Celsius devised a method in which absorbing glass plates were inserted into the optical train of a telescope observing a star. Each plate reduced the brightness of the star by a half-magnitude, so the total number of inserted plates needed to extinguish the star could be used to find that star's magnitude (Hearnshaw, 1996). This principle led to the development of the **wedge photometer**, one of the more common types used in astronomy at that time. In a wedge photometer, a tapered wedge of optically absorbing glass is inserted into the light path; the further it is inserted, the thicker the glass in the path and the greater the attenuation of the star light. The thickness of the glass at the point at which the star is no longer visible can be used to find its brightness (Hearnshaw, 1996).

Although the extinction photometer works well in principle, the eye is more reliable at comparison than detection. The **comparison photometer** takes advantage of this and works by using the eye as a null instrument to compare the brightness of two sources, one of which may be a standard. Figure 1.2 is an adaptation of one of Lambert's figures illustrating an early, very simple comparison photometer that uses the inverse square law to quantitatively compare sources of light. The two sources in the figure consist of single and twin candles; both sources illuminate the wall on the far right of the figure. A standing wall in the middle of the room separates the two illumination sources so that the wall area labeled **A** is only lit by the single candle and the wall area labeled **B** is lit only by the twin candles. The brightness of the two sides of the wall can be compared where they meet in the center, indicated by a vertical dashed line in the figure. When the candles are the same distance from the wall, **A** is darker than **B** because its illumination source is half the other. To find the relative brightness of the two sources, the single candle is moved until the brightness of **A** and **B** are indistinguishable by eye, and the distance of each light source from the far wall is recorded. If, as in this case, the candles are identical, the inverse square law requires the single candle to be the square root of two (0.71) times closer to the wall than the twin candles for the illumination to be the same.

This type of photometer worked well when Lambert was experimentally developing his ideas, but is impractical if portability is needed. In 1794, **Benjamin Thompson**, aka **Count Rumford** (1753–1814), developed the first practical photometer (Thompson, 1794; DiLaura, 2006). His apparatus was essentially a miniature version of Lambert's; the source and a standard light were placed on separate ruled rails and could be moved along them. These sources jointly illuminated a small screen, partially shadowed with obstructions. The user looked at the screen and moved the source in question closer or

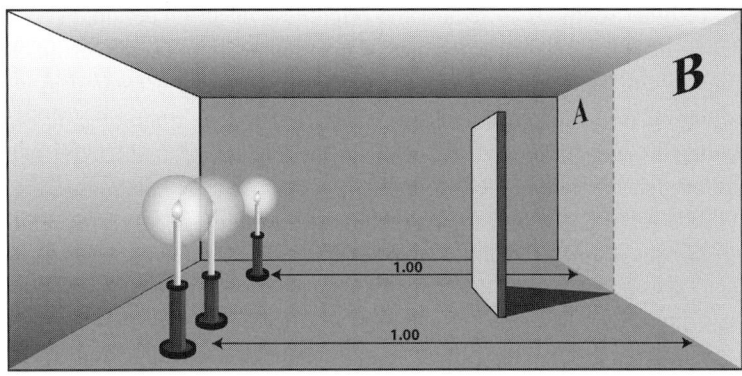

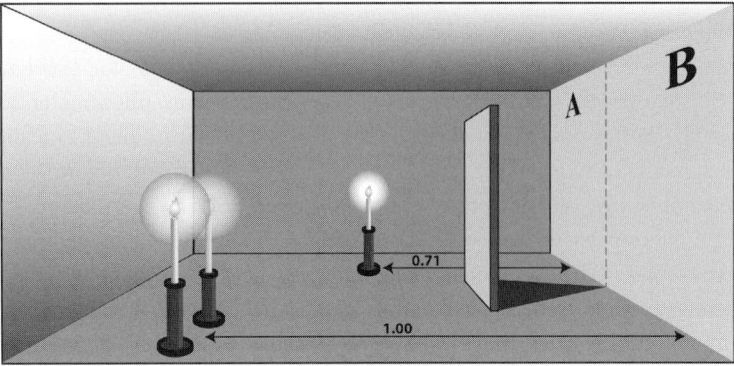

Figure 1.2. An illustration of a comparison photometer, based on a figure in Lambert's Photometria. See the text for details.
Credit: Michael K. Shepard.

farther as necessary to make the screen evenly illuminated. The distances of the source and standard were then read and used to calculate the brightness of the source relative to the standard using Lambert's photometric laws. This early design is often referred to as the **Rumford** or **Rumford-Lambert photometer**.

For astronomy, a different design was necessary because, with the exception of the Sun and Moon, astronomical objects of interest are point sources and extremely faint compared even to a single candle. It's not possible to move them closer or farther and use the inverse square law as in the Rumford–Lambert photometer, so one must develop other methods to reduce their light in a controlled manner.

Using two identical telescopes, **William Herschel** (1738–1822) developed one of the first comparison photometers for two stars. On the telescope observing the brighter of the two stars, he used an aperture stop to reduce the incoming light (like the pupil of your eye) until the two stars were indistinguishable in brightness. The difference in the two apertures could then be used to determine their relative brightness (Hearnshaw, 1996). It was also found to be helpful to compare the brightness of stars out of focus, as these could be compared more readily than point sources.

The pinnacle of photometric instrumentation up until the twentieth century was designed in 1857 by **Johann Karl Friedrich Zöllner** (1834–1882), a German astrophysicist at Leipzig University (Staubermann, 2000; DiLaura, 2001; Figure 1.3). The Zöllner photometer utilized a small telescope and a kerosene lamp flame as the standard – an artificial star. Light from a star or planet entered the telescope directly, while the light from the lamp entered via a reflective wedge in the light train; this allowed both the star and standard to be seen at the same time in the telescope eyepiece. Polarizing Nicol prisms

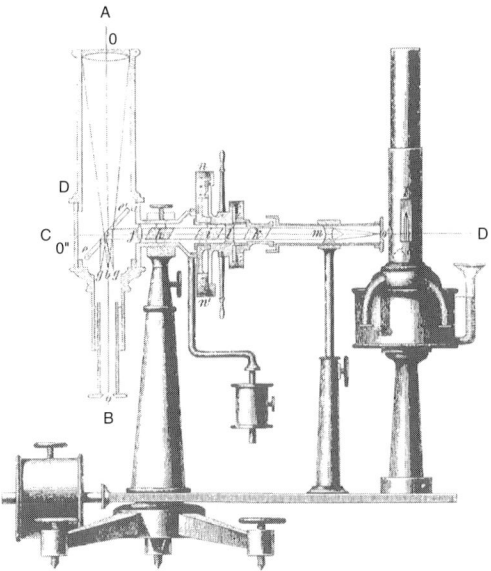

Figure 1.3. A drawing of a Zöllner comparison photometer. The starlight entered through the objective, O, along the optical axis, A–B, and was viewed by the observer at B. The standard kerosene flame was housed in the lamp assembly to the right. Its light passed through crossed polarizing filters (to reduce intensity) and a birefringent material (to change the color) along the axis labeled C–D and was partially reflected from the glass plate at e-e′ before being seen alongside the star by the observer at B. Used with permission from the Royal Observatory, Edinburgh.

were in the optical train of the lamp; when these were rotated relative to each other, they provided a simple and measurable way to dim it. In this way, the light from the star and lamp could be made equal, and the amount of rotation necessary allowed for the calculation of their relative brightness. Later versions included a way to pass the lamp light through a birefringent plate to adjust the flame color; in this way, the astronomer could better estimate the relative brightness of stars with noticeable color (Sidgwick, 1971; Staubermann, 2000; DiLaura, 2001).

Perhaps the most successful comparison photometer of the nineteenth century, in terms of sheer volume of measurements, was a **meridian photometer**, installed by astronomer **Edward Pickering** (1846–1919) at Harvard in 1879. Like Herschel's early photometer, this one used twin telescopes, each with a 4 cm objective; one telescope always pointed to Polaris, the standard, and the other pointed toward any star to be measured, so long as it was on the meridian (Hearnshaw, 1996). This latter condition helped to ensure the loss of light (extinction) from the atmosphere was minimized. Like the Zöllner photometer, the optical path included polarizing filters to reduce the brightness of either star until they were equally bright. With this photometer, Pickering observed more than 4,000 bright stars for the *Harvard Photometry*, one of the first major star magnitude catalogs (Hearnshaw, 1996).

1.5.3 Photometric Standards

For a standard light source other than a star, scientists chose readily available and easily controlled devices. Thompson used a wax candle 0.8 inches in diameter, which burned 109 Troy grains (7 g) of wax per hour, viewed at a distance of 10 inches (25.4 cm) (DiLaura, 2006). Others used candles made from spermaceti – the oil from the sperm whale. For much of the late nineteenth century in Germany, the Hefner lamp, burning amyl acetate, was the standard, and we noted earlier that the Zöllner photometer employed a kerosene lamp (DiLaura, 2001, 2006). The disparity in standards made comparisons between different laboratories or instruments more difficult. In 1909, the first **International candle** was defined and there were several attempts in the twentieth century to standardize a quantity of perceived brightness (luminous intensity, or perceived power by a human eye per solid angle). The most recent definition was adopted in 1979 by the International Bureau of Weights and Measures (BIPM, 1979):

> The **candela** is the luminous intensity, in a given direction, of a source that emits monochromatic radiation of frequency 540×10^{12} Hz and that has a radiant intensity in that direction of 1/683 watts per steradian.

You may have noticed that this is a physical definition, divorced from the older concept of perceived brightness by the human eye. It has been tailored to approximate the earlier standards both by the chosen frequency – a wavelength of 0.555 μm is green light and the peak sensitivity for the human eye – and a radiant intensity that can only have been chosen because it approximated the perceived brightness of a candle. The candela is one of the seven base units in the Système International (SI).

1.5.4 Standard Stars and the Problem of Color

For astronomy, it was often convenient to use another star as a standard. Although the Pogson interval had become the de facto standard, there was still no agreed-upon **zero point**, a reference intensity of magnitude $m = 0.0$, or standard star(s) with which to compare every other star. Competing systems of photometry were devised by groups in the United States (Harvard, under Pickering), England (Oxford, under **Charles Prichard** (1808–1893)), and Germany (Potsdam, under **Gustav Müller** (1851–1925) and **Paul Kempf** (1856–1920)) (Hearnshaw, 1996). They all agreed on the Pogson interval, but differed in their standards.

Pickering had originally adopted Polaris as defining $m = 2.0$. To take into account the extinction caused by the atmosphere, Pickering adopted an extinction coefficient of 0.25 magnitudes per unit air mass (air mass = 1.0 if observing at the zenith. See Chapter 3). He also revised his magnitude value for Polaris to agree with previous estimates. When all was said and done, he *defined* Polaris to have $m = 2.15$ when observed at the zenith (Hearnshaw, 1996). Prichard also adopted Polaris as the standard, and with similar reasoning defined it to have $m = 2.05$, an unfortunate tenth of a magnitude brighter than Pickering (Hearnshaw, 1996). The Potsdam Catalog used a completely different standard system: a group of 144 stars, spaced uniformly around the sky, were defined as standard stars with magnitudes ranging from 4.5 to 7.3 (Potsdam, 1894; Hearnshaw, 1996). The labor involved in initiating this kind of system appears to have paid off in a more precise and internally consistent catalog than the other two (Hearnshaw, 1996).

Polaris was soon found to be variable star and unsuitable for a standard. Today, the zero point for most systems derives from an early (but not current) definition that the A0 star Vega (α Lyrae) has a visual magnitude $m_v = 0.0$ (Bessell, 2005). Modern practice has also adopted the Potsdam method of defining a series of standard stars around the sky, such as the Landolt stars (see Chapter 2). Today, the system is better standardized, if more complex, and there are efforts to once again divorce the concept of magnitude from a particular star to a physically defined standard. We will discuss this more in Chapter 2.

The last major problem to be tackled was color. The eye perceives red and blue differently – how do we compare red and blue stars? Zöllner and others had used methods to manipulate the color of their standard light sources to match that of the target. This, however, was a very subjective correction. Today, astronomers use sets of standardized color filters to restrict the measured light to particular wavelength band passes; in this way, a star or planet may have several different magnitudes, depending upon the color filter used. Again, we defer this discussion until Chapter 2.

1.6 The Brightness and Color of Stars and Planets

A great deal of time was spent in the late nineteenth and early twentieth centuries measuring and rectifying stellar magnitudes into consistent catalog systems. As we've seen, the biggest issues for much of this time were inconsistent standard stars, no agreed-upon zero point, and the problem of color. While these issues were thorns in the side, they could not compare to the difficulties facing scientists attempting to measure the magnitude of the Sun or Moon. And while the brightness of planets could be measured as if they were stars, physical characterization required knowing the brightness of the Sun in order to derive planetary albedos. The best work in this field up until the early twentieth century was discussed in a 1916 paper by one of the deans of modern planetary science, **Henry Norris Russell** (1877–1957), and we summarize it in the following section.

1.6.1 The Sun

The Sun is arguably the most difficult photometric target for astronomers. Its half-degree angular size coupled with its overwhelming brightness means that very creative methods must be devised to compare it with other stars. Russell (1916a) detailed four innovative studies. In 1865, Zöllner, using his eponymous photometer, used a series of lenses and attenuating filters (Russell calls them "shade-glasses") to take the large disk of the Sun and reduce it to a dimmer, point-like source that could be compared with his artificial light source. At night, he used the same apparatus (without the attenuating optics) to measure Capella (α Aurigae). This is actually a quadruple star system, but the brightest pair are G-type stars and similar in visual color to the Sun. Zöllner made 6 measurements of the Sun and 11 of Capella and reported a difference in magnitude of 26.87 between the two. Using the then accepted magnitude (on Pickering's Harvard scale) of +0.21 for Capella, Russell derives a solar magnitude of $m_{sol} = -26.66$.

Near the turn of the twentieth century, **Charles Fabry** (1867–1945) compared the Sun to Vega (α Lyrae) using an electric lamp standard and derived a difference of 26.94 magnitudes between them. On the same Harvard scale, Russell derives $m_{sol} = -26.80$. In 1903, **Vitold Karlovich Ceraski** (1849–1925) compared the Sun with Venus, and then Venus with Procyon (α Can Minor), Polaris (α Ursa Minor), and Sirius (α Can Major). Of course the magnitude of Venus changes from night to night, but accounting for these changes, Russell reduces these observations to $m_{sol} = -26.60$. Lastly, in 1901 Pickering used a shadow photometer and blue-filtered pentane lamp to compare extra-focal images of the Sun and four reference stars to obtain $m_{sol} = -26.83$. These four estimates are remarkably similar despite the various methods, and Russell averages them to find $m_{sol} = -26.72 \pm 0.04$. Although the Harvard scale differs slightly from the modern magnitude system (see Chapter 2), the currently accepted apparent visual magnitude for the Sun is –26.75 (Cox, 2000).

1.6.2 The Planets

Compared to the latter half of the twentieth century, the state of the art in planetary photometry in 1916 was quite primitive. Russell lists all of the studies known to him, and credits Müller (Potsdam) with the best and most complete data set to that point. The overriding issue of the time was acquiring an accurate magnitude for each planet, and secondarily, characterizing how their magnitude changed with movement relative to the Earth and Sun. In Chapter 2, we introduce the concepts of absolute magnitude and phase angle. For now, we note that the absolute magnitude of an object is its magnitude if seen when fully illuminated at a standard distance from the Sun and Earth. Phase angle is measured from the Sun to the object and back to Earth, and for reference, 0° appears fully illuminated, 90° appears half-illuminated, and 180° is not illuminated because it is between the Earth and Sun.

A major difficulty in early measurements of Mercury and Venus was that, because of their interior orbits, they could only be seen when partially lit. In Russell's time, Mercury had only been measured at phase angles of 50°–120°, and Venus from phase angles of 24°–156°. Mercury's brightness with phase showed it to behave similarly to the Moon (see next section), but Venus did not. Extrapolating magnitudes from these phases to 0° involved significant uncertainty.

The outer planets were considerably easier since they could all be observed quite close to 0°. Mars diminished linearly in magnitude with phase angle, as did the four largest asteroids (see Chapter 6). Jupiter, on the other hand, showed almost no variation in brightness from 0° to 11°. Saturn, like Jupiter, also showed little variation in brightness when the ring was not visible (edge on as seen from Earth), but displayed Mars- or Mercury-like drops in brightness with phase when it was visible. Uranus and Neptune were so far away that they showed little change in phase angle or brightness.

1.6.3 The Moon

The object that dominated planetary and photometric studies up until the mid-twentieth century was the Moon. Without a telescope, the surface of the near-side of the Moon is a mottled disk with roughly equal parts of bright and dark terrain. The Italian astronomer **Giovanni Battista Riccioli** (1598–1671) is generally credited with naming the dark areas **maria** (singular **mare**) and the light areas **terrae** (singular terra), which are Latin for "seas" and "lands" respectively (Moore, 1983). With telescopic aid, it became apparent that the maria were not water (the presence of craters within a body of water would be a clue) but the actual nature of these areas remained a source of contention well into the twentieth century.

Though similar in angular size to the Sun, the Moon is considerably dimmer and therefore easier to compare in brightness with stars. It also shows a complete range of phase angles, near 0°–180°, providing an ideal data set to test early light scattering models (Figure 1.4). Russell discusses seven different studies of the lunar phase curve, the earliest being that of Sir **John Herschel** (1792–1871), son of earlier mentioned William Herschel. The seven sets of observations differed based on the magnitude system assumed, but all showed two interesting features.

First, the first quarter Moon was slightly brighter than the third quarter Moon. This was readily explained, though, because inspection showed the third quarter Moon to have a larger areal proportion of the darker mare material. The second feature, and the more mysterious, was the unusual brightening behavior as the Moon becomes full that we now refer to as the **opposition surge**.

Figure 1.5 shows the phase curve of the Moon derived by Russell (1916a). Also shown are two models for light scattering that were known in Russell's time. The Lambertian curve is the expected brightness for a sphere covered by a material that obeys Lambert's cosine emission law (Section 1.4), and the Area law curve assumes the surface brightness is only a function of the visible area of the Moon. If the Moon obeyed the Area law, the full Moon would be

Figure 1.4. The phases of the Moon observable on Earth.
Credit: NASA/JPL-Caltech/Bill Dunford. www.jpl.nasa.gov/edu/teach/activity/moon-phases/

equal in brightness to the sum of the first and third quarter moons (the right and left halves of a half-illuminated Moon), and if it obeyed Lambert's law, it would be a bit more than 1.5 times as bright as this. In reality, the full Moon is *five times* as bright as sum of the first- and third-quarter moons. The resolution of this mystery would not be settled for 50 years.

1.7 What Was Old Is New Again[2]

1.7.1 Race to the Moon

In the late 1950s and early 1960s, the Moon was the target of choice in a space race between the United States (USA) and Union of Soviet Socialist Republics

---

[2] Much of this section is based on an interview the author and P. Helfenstein conducted with B. Hapke in May 2015.

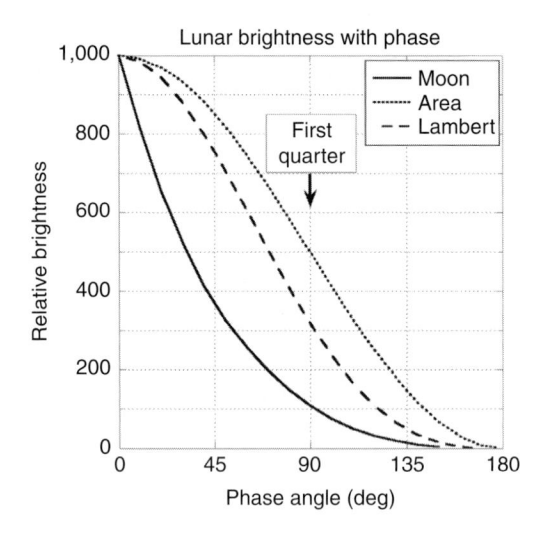

Figure 1.5. Lunar phase curve according to Russell (1916a) along with phase curves for spheres covered with material obeying the Lambertian cosine emission law and the Area law.
Credit: Michael K. Shepard.

(USSR). Both countries had ambitious space programs. They were developing the technology to launch and return humans into space, to orbit the Earth, to spacewalk, to rendezvous and dock spacecraft, and to land on other planets. They also developed, in parallel, sophisticated unmanned spacecraft bristling with cameras and other sensors to flyby, orbit, and land on the Moon and other planets.

As engineers began to plan for unmanned, and later manned, missions to the Moon, it became increasingly important to thoroughly map the surface and determine its physical nature. The maria were obviously smoother than the terrae and the targets of the first unmanned, and later manned missions. But what was the nature of the surface? There were two endmember camps.

On one side, **Gerard Kuiper** (1905–1973), the University of Arizona astronomer sometimes described as the father of planetary science, was convinced that much of the lunar surface was solid lava. He envisioned a surface, in places, rendered into a volcanic "sponge" or "foam" by the constant bombardment of micrometeorites so that it would be porous but solid. At a conference at MIT in 1964, he is also reported to have claimed that, based on the first photos returned from the Ranger 7 spacecraft, "there cannot be a layer of cosmic dust even one millimeter deep on the lunar surface" (Simons, 1964).

On the other side, **Thomas Gold** (1920–2004), the Cornell University astrophysicist and iconoclast, was equally convinced that the Moon was covered with a blanket of dust and that the maria were essentially large ponds created by the slow transport and settling of dust created during impact events (Gold, 1955). Gold was not afraid to make bold (and some would say outrageous) predictions, and based on his intuitive sense that dust would be widespread on the Moon and using the analogy of hidden snow-filled crevasses in the glacial ice-fields, he is known to have warned of similar traps for astronauts. Some have claimed that he warned of a hundred-meter-thick layer of fine dust over the entire Moon that would lead to catastrophic consequences for landing astronauts, but this claim does not appear in print.

How could one differentiate between these two hypotheses without landing on the lunar surface? Gold hired a postdoctoral research associate named **Bruce Hapke** (University of Pittsburgh) to figure it out. Hapke first took on the task of attempting to find a way to transport dust over the surface. Gold had hypothesized that the solar wind might electrostatically charge the dust, allow it to levitate, and thus slowly move along gravitational gradients into lows. Hapke constructed a vacuum chamber to hold dust and a hydrogen-ion gun to bombard and, if Gold was right, charge and levitate the dust. But the experiments failed to achieve this goal. This is still an active area of research interest, and Gold's idea may yet be confirmed in some form. For while we know the maria are not dust ponds, features like this have been observed on asteroid 433 Eros (Robinson et al., 2001).

1.7.2 Fairy Castles

Although the levitation experiments did not confirm (or refute) the prospect of dust on the lunar surface, there were other avenues to test this hypothesis. Throughout the 1960s, scientists employed a variety of remote sensing techniques to investigate the surface conditions in anticipation of the Apollo program. Some looked at the radar reflectance of the Moon (e.g. Evans and Pettengill, 1963; Thompson and Dyce, 1966), while others measured its thermal inertia (e.g. Shorthill and Saari, 1965). But the unusual backscattering behavior of the Moon at opposition was still unexplained. Most common substances – wood, paper, etc. – reflect light in a more-or-less uniform diffuse pattern referred to as Lambertian (see Section 1.4.2), but as shown by Russell and others before, the Moon does not. As far as anyone knew, this behavior was unique to the Moon. Gold was convinced it was caused by a ubiquitous layer of fine dust and encouraged Hapke, with the help of then graduate student **Hugh van Horn** (University of Rochester), to also pursue this line of work in the laboratory.

To measure the photometric behavior of a sample in the laboratory, one needs a device that can illuminate the sample with a filtered and collimated beam of light that can be variably oriented. To be most effective, it needs to illuminate the sample from directly overhead to near-grazing angles. One also needs a detector configured to measure the reflected light from an equally wide range of angles. A device capable of illuminating and measuring samples in this way is called a **photometric goniometer**, from the Greek for angle and measurement (see Chapter 7).

After constructing their goniometer, Hapke and van Horn spent months measuring the scattering properties of some 200 materials. Because the measurements require illumination from a range of directions and measurement from a variety of directions, the collective scattering behavior of a material is referred to as its **bidirectional reflectance distribution function** (BRDF). There are a number of ways to represent the angular scattering behavior of a surface, and technically, the BRDF is one very specific way (Chapter 3). However, the term is often used generically to mean the overall scattering behavior of a surface, and it is in this sense that we use it in this chapter.

Figure 1.6 illustrates the BRDF of a model Lambertian surface, a sample of common sand, and lunar regolith returned from Apollo 11. The plot shows the apparent brightness (what we will call the **radiance factor** in Chapter 2) of the scattered light (y-axis) versus the viewing (emission) angle at which the measurements are made (x-axis). In this case, the samples are illuminated from an angle 45° from vertical (incidence angle = 45°). In both experiments, the viewing direction is varied from directly overhead (emission angle = 0°) to grazing (emission angle = ±60°) in both the forward (negative emission angles) and backscattering directions (positive emission angles).

In the particular way these plots are made, a Lambertian surface looks uniformly bright at all viewing angles. And although there are some differences, common sand scatters light in approximately the same way. The lunar regolith, however, is distinctly different. It is darker at viewing angles in the forward scattering direction, and gets significantly brighter as the illumination and viewing angles begin to coincide in the backscattering direction. This is the opposition surge (Section 1.6; see also Chapter 5).

In the early 1960s, Hapke and van Horn did not have the benefit of real Moon dust to test, and they were having difficulties observing this scattering behavior in the laboratory. None of the surface materials proposed by scientists displayed the observed lunar scattering behavior; they tried rock,

1.7 What Was Old Is New Again

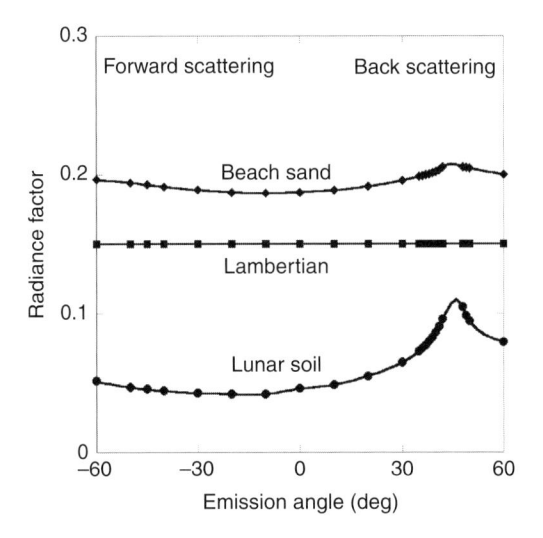

Figure 1.6. Bidirectional reflectance curve (radiance factor) of a model Lambertian surface, a measured sample of beach sand, and a measured sample of lunar soil (10084). All samples are illuminated by collimated light 45° from the zenith. The y-axis is the normalized brightness and the x-axis is the emission angle – the viewing angle of the observer measured from the zenith. Note that the Lambertian surface looks equally bright from all viewing directions, the beach sand is similar, but the lunar soil shows a very sharp backscattering behavior or opposition surge. Credit: Michael K Shepard.

crushed rock, pumice, obsidian, glass beads, and even metal powders. They only began to see hints of the behavior when they looked at very fine-grained (8 μm) silicon carbide, a dark crystalline powder used as an abrasive. Then they got creative with sample choice and began to see the behavior in unexpected materials – lichen, grass, and a piece of sponge. A few years earlier, van Diggelen (1959) had also noted that moss exhibits the same type of photometric behavior. These findings suggested a very intricate material.

After many additional experiments, the strong backscattering behavior was isolated to the more geologically realistic combination of extremely fine-grained, dark samples, in a highly porous and interconnected configuration. Hapke and van Horn coined the term **fairy castle structure** to describe this configuration and estimated that the media had to have a porosity of about 90%! These conditions could only be achieved with micrometer-scale particles in which surface adhesive forces overcome gravity and allow dizzying edifices to exist in the upper few millimeters of the powdered surface.

1.7.3 The Dark of the Moon

To achieve the backscattering effect, the darkness or low **albedo**, of the samples was just as important as their fine granularity, for the effect appeared to be caused by self-shadowing. In bright samples, shadows are largely washed out. But here was a conundrum: when silicate rocks are ground into finer and finer sizes, they become brighter. Yet the lab experiments suggested both very fine *and* dark grains were required.

From his previous work bombarding dust samples with simulated solar wind, Hapke (1962) had noticed that irradiated samples were darkened by some unknown mechanism. Similar effects had been noticed by Wehner et al. (1963) and they and Hapke (1962, 1965) suggested that the newly discovered solar wind could be the cause of the darkened lunar surface. It would be nearly 40 more years before the mechanisms for darkening were experimentally verified, but we now know that micrometeorite bombardment and sputtering induced by the action of solar wind form micron-scale agglutinates consisting of rock fragments and dark glass, vapor-coated with nanoscale iron (Keller and McKay, 1993; Hapke, 2001; Sasaki et al., 2003). These processes work to both darken and redden the lunar **regolith** and are now collectively referred to as **space weathering**.

1.7.4 Ground Truth

On February 3, 1966 – 50 years after Russell published the odd phase curve of the Moon – the Soviet Union's Luna 9 mission successfully soft-landed on the Moon. The US Surveyor 1 followed 4 months later. These and follow-on unmanned and manned missions confirmed what Gold had suspected, and Hapke and van Horn had experimentally demonstrated (Figure 1.7). The Moon was covered with a layer of extremely fine-grained dust (or **regolith**), several centimeters deep. The fears attributed to Gold – that the dust layer was tens to hundreds of meters in depth – were largely assuaged. The same meteorite bombardment that created the ubiquitous dust also compressed the subsurface to a dense substrate. In fact, the Apollo astronauts on all missions had great difficulty in acquiring soil cores even a meter deep (Heiken et al., 1991).

The US and USSR space programs were massive and expensive undertakings, and provided scientists with a still glittering treasure trove of return samples to study. And yet, for a minuscule fraction of the cost of those programs, scientists on Earth were able to surmise a great deal about the state of the lunar surface without ever leaving the laboratory or observatory.

1.7 What Was Old Is New Again

Figure 1.7. Surveyor 3 and Apollo 12 astronaut Pete Conrad.
Credit: NASA, Alan Bean. www.hq.nasa.gov/alsj/a12/images12.html, photo AS12-48-7134HR.

The field of planetary photometry originated centuries ago when we could only look with some longing at the far-away places in our solar system. Although it undergirds the calibration of all imaging sensors, the glamour of this patrician field has waned in the past decades as spacecraft have brought us up close and personal. But now, a century after Russell, we have discovered thousands of new planets around distant stars, and today's extra-solar planetary scientists find themselves in much the same place as their progenitors more than a century ago – straining to learn as much as possible from a few photons reflected from these distant places.

2
Photometry Conventions, Terminology, and Standards

In this chapter, we examine the conventions of coordinate systems and time, before describing the major terms, like *radiance* and *irradiance*, used in planetary photometry. Then we introduce the magnitude scale and look at the most common photometric systems used in planetary photometry, and their standards for calibration.

2.1 Coordinate Systems

Planetary photometry was born from the perspective of Earth. We orbit the Sun at a distance of 1 astronomical unit (AU) and, from this perch, view all the other planets as they orbit the Sun in their respective orbits. Therefore, most of the terms we use and observations we make are from this geocentric and historical perspective. With the advent of the space age, robotic emissaries allowed much closer views of the planets, often in viewing conditions impossible to obtain from Earth. However, the terminology used is the same and, for convenience in use, the spacecraft becomes Earth and our terms shift to this new frame of reference accordingly.

To locate an object in space requires at least four coordinates: three spatial and one temporal. Traditionally, the spatial coordinates have been based on the spin axis of the Earth and its equator. Time has been based on the length of the day and year. As instrumentation became more sophisticated, these reference systems required more exact specifications. In the next sections, we provide a general overview of the spatial reference systems used by astronomers before discussing the two most likely to be encountered by planetary scientists: **equatorial** and **ecliptic**.

2.1.1 International Celestial Reference System and Frame

Most of us drive an automobile, but few of us understand how it works beyond "insert key, start, and go." Engineers design the car and mechanics service them, but their work is largely invisible to us. Similarly, the common reference frames used by astronomers and planetary scientists are relatively easy to understand and use, but have a background infrastructure and periodic servicing that are largely hidden to us. There is an extensive literature on astronomical reference frames and time keeping, and an excellent introduction with references to further reading can be found in Kaplan (2005). Here, we provide only the briefest of overviews so that the reader can glimpse a little of what goes on behind the scenes.

A **reference system** is the set of *rules* used to define a **reference frame**. The reference system will include definitions of the origins and axes of the system and any constants and formulae needed to convert observable objects into this system. The reference frame is the practical *realization* of this system and generally consists of a number of observable points in the sky and their coordinates in the specified system (Kaplan, 2005).

In an ideal world, the reference frame will be fixed and unmoving. Unfortunately, this is not possible to achieve in practice. All of the stars in our galaxy appear to move over timescales of decades to centuries. Fortunately, there are objects well beyond our galaxy that are, for practical purposes, essentially fixed. With this in mind, the International Astronomical Union (IAU) developed **The International Celestial Reference System (ICRS)**. In 1991, the IAU (1992) recommended that this reference system be defined such that

> the space coordinate grids with origins at the solar system barycentre and at the centre of mass of the Earth show no global rotation with respect to a set of distant extragalactic objects.

The center of mass, or **barycenter**, of the solar system is near the center of the Sun, but not at the exact center because of other mass in the solar system. It was also urged that, for continuity, the new reference system align closely with the major reference already in existence: the J2000.0 equinox (see next section). In 2000 and 2006, the IAU followed up with definitions of the **Barycentric Celestial Reference System (BCRS)** and essentially made it equivalent to the ICRS (IAU, 2001, 2008).

The realization of the ICRS adopted by the IAU (and currently in use worldwide) is the **International Celestial Reference Frame (ICRF2)**, which uses a catalog of more than 200 primary and some 3,000 other secondary extra-galactic radio sources as "fixed" points. The reasoning is that these objects are so far away that, over timescales relevant to most applications, they will not appear to move. Their positions are measured using **Very Long Baseline Interferometry (VLBI)**, and they have been chosen such that systematic and random positional errors do not exceed 0.1 milliarcseconds. A second realization is the **Hipparchus Celestial Reference Frame (HCRF)**, which is the optical equivalent of the ICRF. However, VLBI is capable of much higher resolution so the HCRF is roughly an order of magnitude less precise (Kaplan, 2005).

2.1.2 Fixed Coordinate Systems

In astronomy and planetary science, we commonly map objects onto the sky using one of two fixed coordinate systems: **equatorial** or **ecliptic** (Figure 2.1). (There are others, most notably the *galactic coordinate* system, but these are little used by planetary science so we ignore them here.) Both of these systems

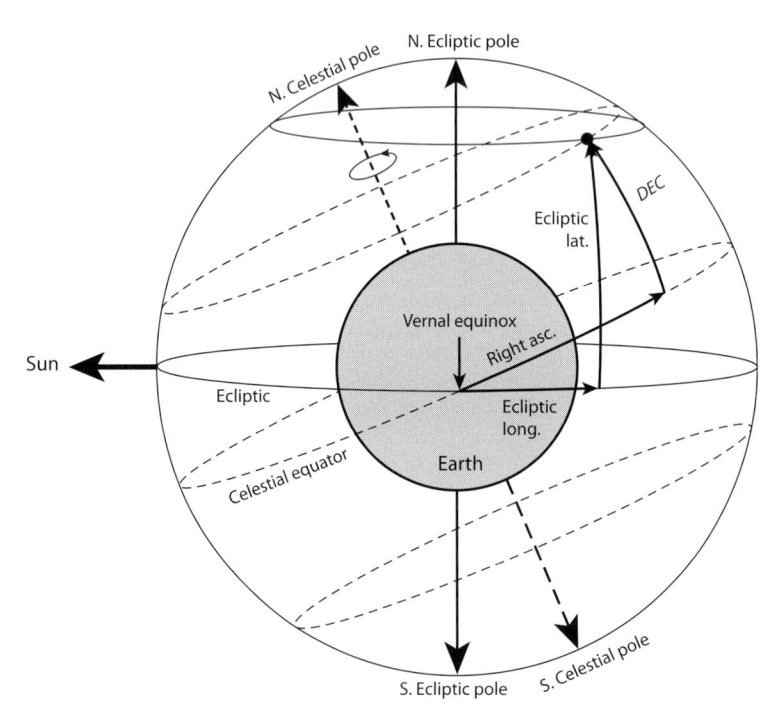

Figure 2.1. The Earth and two coordinate systems: equatorial and ecliptic. The Earth spins about an axis pointing to the N. and S. Celestial Poles. The ecliptic is defined as the path the Sun appears to take in the sky. An object of interest, shown as a small dot, can be referenced to either system: equatorial uses the right ascension and declination, while ecliptic uses the Ecliptic longitude and latitude. Credit: Michael K. Shepard.

2.1 Coordinate Systems

are tied to the ICRF but, unlike it, neither is truly fixed; they shift on decade-long timescales that require minor adjustments.[1]

The equatorial system is most often used by astronomers and planetary scientists when referring to the locations of objects relative to a star catalog; it is the system used by telescopes to point and track our targets. The ecliptic system is most often used in planetary science as a way of orienting features with respect to the plane of the solar system, defined by the orbit of the Earth around the Sun, and approximated by the orbits of everything else between the Sun and the Kuiper Belt. For example, the directions of the rotation axes of planets and asteroids are often reported in the ecliptic reference system because this gives one an immediate picture of how the spin axis is oriented with respect to the ecliptic plane of the solar system. Because most asteroids orbit within this plane (or near enough), it also gives a sense of how the spin axis compares with the objects direction of motion (e.g. is it oriented "upright" or tilted over?).

To define the **equatorial coordinate system**, we project the Earth's equator and spin poles onto the sky to define a **celestial equator** and **north and south celestial poles**. In analogy with the Earth's latitude system, we designate the position of an object north or south of the celestial equator as **declination**, abbreviated as ***DEC*** or δ. We will use ***DEC*** in this book. The celestial equator is at ***DEC*** $= 0°$; the north and south celestial poles are at $+90°$ and $-90°$, respectively. The position east and west, analogous to longitude in the geographic coordinate system, is called **right ascension** and abbreviated as *RA*, α, or *h*. We will use ***RA*** in this book. It is measured east from the **vernal equinox**, the point in the sky where the Sun crosses the celestial equator on the first day of spring. Right ascension is often measured from 0 to 24 h, a historical artifact from the way in which it was originally determined. Alternatively, one may measure it in degrees, $0°-360°$, recognizing that each hour of ***RA*** equals $15°$. When combined with the sidereal time, right ascension tells astronomers whether an object is "up" above the horizon and in position for observation.

A complication with this system arises because, like a spinning top, the Earth's polar axis wobbles or **precesses** with a period of 26,000 years, slowly moving the direction of the vernal equinox about $0.014°$ (20 arcseconds) per year. In addition, there is another set of smaller periodic epicycles of

---

[1] When the ICRS was implemented, the standards for defining reference positions and planes changed. The newer system was aligned as much as possible with the older systems described here, but newer concepts and terms were employed. For example, in the ICRS "Celestial Ephemeris Origin" replaces "equinox," "Earth Rotation Angle" replaces "Greenwich Sidereal Time," "Celestial Intermediate Pole" replaces "north celestial pole," etc. Nevertheless, the historic or classical terms, while technically deprecated, are still widely used as we do here.

motion about the main direction of precession; these are collectively referred to as **nutation**. Finally, many stars have measurable motion from year to year – referred to as **proper motion** – and this must be considered. Without pinning the equatorial coordinate system to a reference time or standard **epoch**, the *RA* and *DEC* coordinates of all stars would be continually changing as a result of these factors. The epoch **J2000.0** is currently the standard reference time, and indicates the orientation of Earth in space as it was on (Gregorian date) January 1, 2000 at 12:00 noon, and all modern observations must take into account precession and nutation to be cast into this frame.

Because the Earth's axis is tilted some 23° off the perpendicular of its orbit around the Sun, it is often more convenient to use a coordinate system tied to our orbital plane. This is the **ecliptic coordinate system**; the name comes from the **ecliptic** – the apparent path of the Sun through the sky. The **ecliptic latitude**, β, is defined as the number of degrees north or south of the ecliptic. The **ecliptic longitude**, λ, is defined as the number of degrees east from the vernal equinox, just as with the equatorial system.

2.1.3 Relative Coordinates

As an alternative (or in addition to) using fixed coordinates, it is common to express the position of a planet with respect to the Earth and Sun. These relationships are useful because they can be directly related to the apparent phase of an object, e.g. crescent phase. For example, when an outer planet is directly opposite the Sun from the Earth, we say it is at **opposition**; at that point, it would appear to be fully illuminated. We have already noted that the brightness of many objects increases disproportionately as they move through opposition, the phenomenon known as the opposition surge. When an outer planet is on the opposite side of the Sun from the Earth (and therefore lined up with it), we say it is at **conjunction**. Interior planets are also said to be at **conjunction** when aligned with the Earth and Sun; **inferior conjunction** if between the Earth and Sun, **superior conjunction** if on the opposite side of the Sun.

A line from the Sun to the Earth to the object (S-E-O) makes an angle called the **solar elongation**, or more simply, **elongation**, ε (Figure 2.2). It is the apparent angular distance between the Sun and planet and is the equivalent of the great-circle shortest distance between two points on the sky. Given the right ascension (in degrees) and declination of the Sun and object (or their equivalent ecliptic longitude and latitude), it can be found from

$$\varepsilon = \cos^{-1}\left[\sin DEC_{Sun} \sin DEC_{Obj} + \cos DEC_{Sun} \cos DEC_{Obj} \cos(\Delta RA)\right] \quad 2.1$$

where ΔRA is the difference in right ascension between the two objects.

2.1 Coordinate Systems

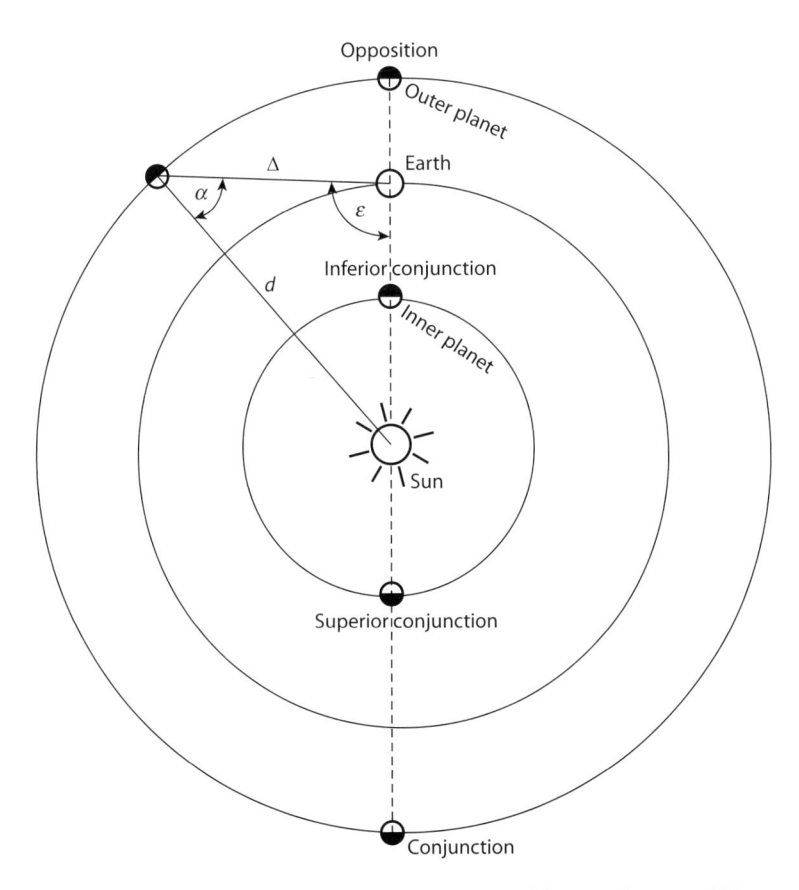

Figure 2.2. Relative positions of the Sun, Earth, and inner and outer orbiting planets, along with the terms for special positions and position angles illustrated. Credit: Michael K. Shepard.

The angle along a line from the Earth to the object to the Sun (E-O-S) is called the **phase angle, *a***. Using the Moon as an example, the phase angle of the Moon at first or third quarters is 90°; at full Moon it is near 0° (opposition); and at new Moon it is near 180°. The phase angle can be readily calculated if one knows the object's distance from the Sun (d) and Earth (Δ) (here in astronomical units):

$$a = \cos^{-1}\left(\frac{d^2 + \Delta^2 - 1}{2d\Delta}\right) \qquad 2.2$$

Technically, opposition occurs when the right ascension of the Sun is 180° (12 h) away from the object in question. This does NOT necessarily mean it is at a phase angle of $\alpha = 0°$ because the object may be outside the plane of the ecliptic. A phase angle of $\alpha = 0°$ requires an object to be 180° away from the Sun in right ascension and to fall on the ecliptic from the perspective of the Earth. In terms of ecliptic coordinates, a phase angle of $\alpha = 0°$ requires the object to be 180° away from the Sun in ecliptic longitude and to have an ecliptic latitude of 0°.

An object with an orbit interior to the Earth can be observed at all phase angles except those when it is directly between the Earth and Sun (inferior conjunction, $\alpha \sim 180°$) or directly behind the Sun (superior conjunction, $\alpha \sim 0°$). However, outer planets with roughly circular orbits have more restricted viewing geometries. They can be observed from $\alpha = 0°$ (opposition) to a maximum phase angle, α_{max}, of

$$\alpha_{max} \cong 2\sin^{-1}\left(\frac{1}{2a}\right) \qquad 2.3$$

where a is the semi-major axis, in AU, of the object in question. So, for example, an asteroid with a semi-major axis of 2.8 AU would have a maximum phase angle of $\alpha_{max} \sim 21°$ as seen from the Earth, while satellites of Jupiter at 5.2 AU would have a maximum phase angle of $\alpha_{max} \sim 11°$. Greater phase angle coverage requires observations by spacecraft or a close approach to Earth (e.g. Near-Earth asteroids or comets).

2.1.4 Surface and Planetary Coordinates

When we are fortunate enough to have resolved images of a planet, either from high telescopic magnification or spacecraft imagery, we must use additional coordinates to specify the photometric geometry. There are two coordinate systems commonly used: surface coordinates and luminance coordinates. **Surface coordinates** are used when the surface is readily resolved and small relatively flat pieces are considered in isolation. **Luminance coordinates** are the equivalent of latitude and longitude, but with reference lines (meridians and parallels) dictated by the source and viewer. They are used when a more-or-less spherical planet is considered as a whole and allow for the comparison of planetary-scale areas at identical viewing or illumination geometry.

2.1.4.1 Surface Coordinates

If the surface can be resolved, three angles must be specified to completely characterize the photometric geometry of the scene: (1) incidence angle; (2) emission angle; and (3) solar azimuth or phase angle (Figure 2.3).

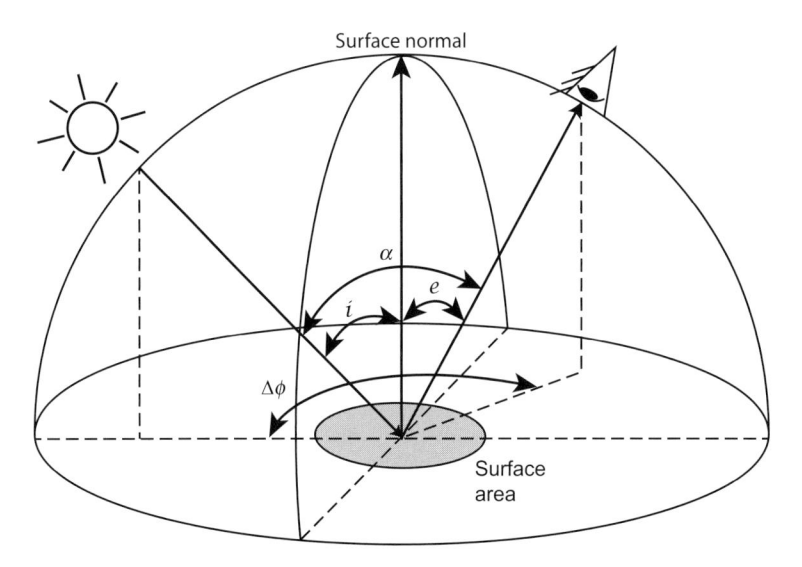

Figure 2.3. Illustration of the viewing and illumination angles used in photometry when the surface can be resolved.
Credit: Michael K. Shepard.

The **incidence angle**, i, is defined as the angle between the surface normal and the vector from the source of illumination (incidence vector). The **emission angle**, e, is similarly defined as the angle between the surface normal and a vector to the observer (emission vector). The solar **azimuth angle** is the angle between the projection of the incidence vector and emission vector onto the surface; because it is a relative measure between two directions, it is usually given as the difference in the two absolute azimuths, or $\Delta\phi = |\phi_{inc} - \phi_{emis}|$. The more commonly used **phase angle**, α, is the angle between the incidence and emission vectors. The solar azimuth and phase angles are related by

$$\cos\alpha = \cos i \cos e + \sin i \sin e \cos(\Delta\phi) \qquad 2.4$$

2.1.4.2 Luminance Coordinates

As with many quantities defined in this field, different authors adopt slightly different conventions for luminance coordinates. We have adopted the conventions used by Hapke (2012).

Imagine illuminating and viewing a spherical planet from a distance many times greater than its size (Figure 2.4). The Sun is coming from the direction of the bold arrow at left, while the observer is viewing from the direction of the

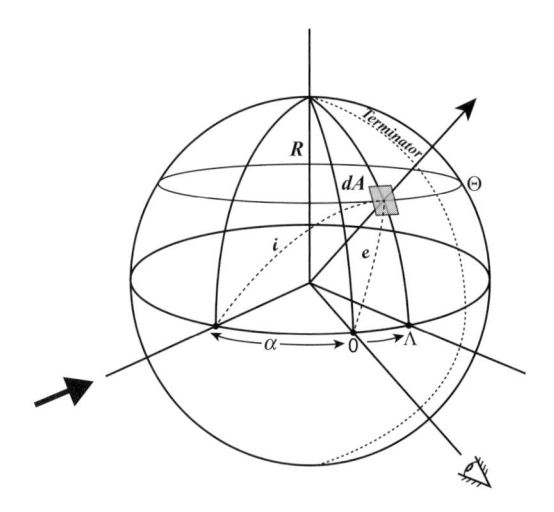

Figure 2.4. Illustration of the luminance coordinate system.
Credit: After Hapke (2012).

eye. Where the sunlight is directly overhead on the planet is the **sub-solar point**; the middle of the observer's view is the **sub-observer point**. A great circle drawn between these two points *defines* the **luminance** or **photometric equator**. By definition, the sub-observer point is on the meridian, 0° longitude (Λ), and 0° latitude (Θ). The sub-solar point is at (Λ,Θ) ($-\alpha$, 0°). The **terminator**, the boundary between light and shadow, is $\pm 90°$ away from the sub-solar point ($\Lambda = \pm 90° - \alpha$) and one side is shown with the dashed line.

What is the illumination and viewing geometry for any given part of the surface? Consider the small gray patch shown in Figure 2.4. Its surface normal is shown with the arrow, and its coordinates are (Λ,Θ). The Sun (or other source) is at an incidence angle, i, from the surface normal, and the observer is viewing it from an emission angle, e, from the normal. These are shown as labeled dashed lines. The angle between these two directions is the phase angle, α.

Most of the scattering or reflectance models we will discuss are in terms of incidence, emission, and phase angles, so we need to relate the luminance coordinates of every point to its illumination and viewing geometry. The Law of Cosines for spherical geometry gives us

$$\cos i = \cos(\alpha + \Lambda)\cos\Theta + \sin(\alpha + \Lambda)\sin\Theta\cos 90° = \cos(\alpha + \Lambda)\cos\Theta \quad 2.5$$

and

$$\cos e = \cos\Lambda\cos\Theta + \sin\Lambda\sin\Theta\cos 90° = \cos\Lambda\cos\Theta \quad 2.6$$

When viewing a planet from a great distance, every point on the surface has the same phase angle, α; only the incidence and emission angle change as you move around. Lines of identical incidence or emission angle are small circles of increasing diameter centered on the sub-solar or sub-observer point, respectively; you may think of them as the patterns on a target "bulls-eye."

As we will continue to point out, different conventions are sometimes used and Equations 2.5 and 2.6 are based on the conventions shown previously. Some authors (e.g. Lester et al., 1979) use the angle from the N. Pole as the "latitude" instead of from the equator as we have, or define the longitude from the point at which the incident vector intersects the luminance equator instead of the emission vector.

2.2 Time

Earlier, we noted that locating the position of an object requires three spatial coordinates (sometimes collapsed into two angles, such as *RA* and *DEC*, when referring to objects in the "plane-of-sky," with the distance from the observer being ignored) and one time coordinate. In recent years, the issue of time has become more complex in astronomy because we are capable of measuring phenomena at temporal scales in which the effects of general relativity become noticeable and important. As with coordinate reference systems, there is an extensive literature on time that we will only briefly consider. Interesting historical background in this field for astronomy can be found in Markowitz et al. (1958), Standish (1998), and Kaplan (2005).

2.2.1 Solar Time

There are two natural temporal cycles that have been historically important in astronomy: the day and the year. For convenience, we define the **local meridian** as the imaginary line that runs from the horizon due north, overhead through the zenith, to the opposite horizon due south. **Local noon** is the time when the Sun crosses the local meridian. The solar day can then be defined as the length of time it takes for the Sun to twice cross the meridian. But because the Earth's orbit around the Sun is elliptical, this interval of time varies slightly over the course of the year. Until industrialization, this was generally unimportant, but with the need to standardize time came the need to standardize the length of the day. Today, the **mean solar day** is defined to be the average time for the Sun to twice cross the meridian. Unfortunately for purposes of standardization, the Earth's rotation rate is slowing, a reaction to tidal braking forces of the Moon, and the length of the mean solar day is slowly increasing.

2.2.2 Universal Time

In 1967, the International System of Units (SI) defined the second as a specified number of vibrations of a Cesium 133 atom as it transitioned between two levels of its ground state. This was later refined to take into account the effects of general relativity and temperature, and today, atomic clocks are keeping time worldwide with a precision of approximately one part in 10^{10}. The time reported by these clocks is referred to as **International Atomic Time** (*TAI*; Kaplan, 2005). Under this system, a **standard day** is defined to be 86,400 SI seconds, divorcing it from the mean solar day.

For historical reasons, we define standard time on the Earth using the relative position of the prime meridian (0° longitude) and the Sun. In the past, this solar timescale was referred to as **Greenwich Mean Time** (GMT) and, when time zones were established, they were referenced to it. In 1928, the IAU replaced GMT with the Universal Time (UT). Although GMT and UT were originally defined the same way (IAU, 1928), Universal Time has undergone several revisions since then, and today GMT no longer refers to any modern timescale; it is only the name of a time zone, much like Eastern Standard Time (EST).

There are several types of Universal Time, but only two are of interest here. **UT1** is defined as the mean solar time computed from the rotation of the Earth within the ICRF and could be considered an *analog* time. **Coordinated Universal Time** (*UTC*) is probably the more familiar Universal Time as it is transmitted by the US National Institutes of Standards on a shortwave radio station (WWV). *UTC* ticks in lockstep with the International Atomic Time and could be considered a form of *digital* time (Kaplan, 2005).

But here is a problem. UT1 is tied to the rotation of the Earth, and it is slowing. *UTC* is tied to a uniformly ticking atomic clock, so it must be periodically adjusted by adding leap seconds if it is to stay in sync with UT1 and the apparent position of the Sun. This is not without controversy, for an enormous amount of our modern infrastructure relies upon precise time keeping. Inserting seconds into the official time is troublesome and expensive. As of July 1, 2015, *TAI* is 36 s ahead of *UTC*.

$$TAI = UTC + 36 \text{ s (as of July 1, 2015)} \qquad 2.7$$

Two other timescales may be commonly encountered. **Terrestrial Time** (*TT*) was defined by the IAU to be an astronomical timescale used to make measurements of objects from the surface of the Earth. It uses the SI second, but is offset from International Atomic Time by 32.184 s to keep it continuous with older dynamical timescales (Kaplan, 2005).

$$TT = TAI + 32.184 \text{ s} \qquad 2.8$$

Ephemerides of solar system objects in the **Astronomical Almanac** (jointly published by the US Naval Observatory, Her Majesty's Nautical Almanac Office, and the UK Hydrographic Office) are listed in Terrestrial Time. Since it is tied to the *TAI*, it is also offset from *UTC* by the same amount as *TAI*.

The SI second is defined within the Earth's gravity field (technically at the surface of the geoid) as it orbits the Sun. Because of the effects of general relativity, clocks on the Earth run slightly slower than those outside of its gravitational well, and our eccentric orbit creates measurable and periodic time-varying dilations. For calculating ephemerides of objects within the solar system, we require a more uniform time standard. In 2006, the IAU defined **Barycentric Dynamical Time (TDB)** for this purpose; this timescale provides a definition of "second" valid for an un-accelerated reference point corresponding to the solar system barycenter (IAU, 2008). This timescale is internally used by organizations maintaining the most accurate ephemerides for the solar system, and then later converted into *TT* or *UTC* using the appropriate relationships.

2.2.3 Julian Day

To specify the date of an event in modern society, we generally use the Gregorian calendar system, e.g. June 30, 2015. But this calendar, while well suited for holidays and seasons, is poorly suited for many astronomical applications. For example, if the rotation period of an asteroid is 9h 4m 3s, and it was observed at midnight on both January 16, 1912 and June 5, 2016, how many rotations took place in that interval? The difficulty comes in attempting to calculate the time interval. It's roughly 104.5 years, but much better accuracy is needed. Years are usually 365 days, but 366 every fourth year. Months have 28, 29, 30, and 31 days. There are also the historical vagaries of the calendar system, such as the 10+ day gap in the civil calendar that occurred when the Gregorian calendar reform was instituted in the sixteenth–twentieth centuries (depending on the country). Calculating the exact interval of time between two dates is awkward and prone to mistakes.

To make this process easier, astronomers use the **Julian Day (JD) (or Julian Day Number**, *JDN*), a running total of solar days elapsed since noon (GMT) on January 1, 4713 BCE in the proleptic Julian calendar or November 24, 4714 BCE in the proleptic Gregorian calendar (Dershowitz and Reingold, 2008).

Here the term **proleptic** simply means each calendar has been extended to dates earlier than its origination. When the Julian Day was suggested by **Joseph Scaliger** (1540–1609) in 1583, Europe used three calendric cycles: a 15-year *indiction* cycle, used for taxation and other civil purposes; a 19-year *Metonic* cycle, the smallest integer number of years with an integer number of lunar months; and the 28-year *solar cycle* (not the sunspot cycle), the number of years between repeats of date and day of the week. By working backward from their values at the time, Scaliger found that Day 1 of each cycle corresponded on January 1, 4713 BCE (Dershowitz and Reingold, 2008).

Julian Days begin at noon and are fractional; for example, the JD for midnight *UTC* January 1, 2015 is 2457023.50, and noon on that day is 2457024.00. The algorithms for converting from a calendar date into JD can be found in a variety of references (e.g. Meeus, 2000) and there are numerous web-based tools available such as those at the US Naval Observatory (aa.usno. navy.mil). Given two Julian Day Numbers, it is straightforward to compute the time difference in days, hours, minutes, and seconds as desired.

One disadvantage of using a Julian Day is that it is seven-digits long and can be unwieldy. A common practice is to truncate the date to five digits. The **Reduced Julian Day (*RJD*)** is

$$RJD = JDN - 2400000 \qquad 2.9$$

and the **Modified Julian Day (*MJD*)** (so that days start at midnight) is

$$MJD = JDN - 2400000.5 \qquad 2.10$$

These abbreviated forms will work until August 31, 2132 CE when the JD will roll over to 2500000 and either six digits will be required or another truncated form will be introduced. Julian Days are often reported in the Universal Time (*UTC*) scale, but for different applications may be in Terrestrial Time (*TT*) or Barycentric Dynamical Time (TDB) scales, in which case there is a difference of several tens of seconds with *UTC*.

2.2.4 Sidereal Time

If instead of marking the passage of the Sun crossing the meridian, we mark that of a star, we find that time interval to be approximately 23h 56m 4s, some 4 minutes shorter than the solar day. This is the **sidereal day**, and the time difference is due to the Earth's roughly 1° orbital shift about the Sun every day, which creates a parallactic shift of the Sun against the far more distant stars. The sidereal day is based on our rotation relative to the Vernal Equinox, which is *defined* to have a right ascension of 0 h.

When the Vernal Equinox crosses the local meridian, the **Local Apparent Sidereal Time** (*LAST*) (or sometimes shortened to **Local Sidereal Time, *LST***) is defined to be 00:00. As the Earth continues to rotate past this point, the *LST* is equivalent to the right ascension of objects crossing the meridian at that moment. So, for example, if the red supergiant star Betelgeuse is crossing the meridian, the *LST* is 05:55:10. Observatories keep clocks with local apparent sidereal time because it is easy to tell, at a glance, the *RA* of objects currently on the meridian. This also happens to be when those objects are at their highest and best placed for observing.

One can determine the local sidereal time simply from the *RA* of objects on the meridian, as above, or it can be determined by a knowledge of the **Greenwich Apparent Sidereal Time** (*GAST* or *GST*) and the local longitude, *lon*. Each 15° of longitude is equivalent to 1 h of *RA*, so

$$LST = GST + \frac{lon}{15°} \text{ (modulo 24 h)} \qquad 2.11$$

where longitude is measured eastward from 0°. For example, 90° W = 270° E. The **local hour angle** (*LHA*) is the angle between *LST* and the *RA* of an object measured westward from the meridian, so it gives one a measure of how far an object is from the meridian. It can be found from

$$LHA = LST - RA \qquad 2.12$$

If the *LHA* is positive, the object is to the west of the meridian; if negative, it is to the east.

Figure 2.5 shows an illustration of these terms. The Earth is shown from the North Pole, rotating counter-clockwise (west to east). The prime meridian (*lon* = 0°) is shown. The observer is at *lon* = 90° W (270° E) and the object of interest is at *RA* = 6 h = 90°. Observers at *lon* = 120° E (240° W) have a local sidereal time of *LST* = 0 h; the Vernal Equinox is on their meridian. For the observers at *lon* = 90° W the local sidereal time is *LST* = 10 h (150° of rotation have occurred since the Vernal Equinox was on their meridian). The Greenwich sidereal time is 16 h. Checking Equation 2.11, we find *LST* = 16 h + 18 h = 34 h mod 24 h = 10 h. The local hour angle of the object of interest is 10 h − 6 h = 4 h, or 60° west of the meridian.

2.2.5 Light-Corrected Time

As related in Chapter 1, Ole Roemer realized that light moved at a finite speed because a periodic event – the disappearance of Jupiter's moon Io – seemed to display different periods. The variation in the period was a result of Jupiter

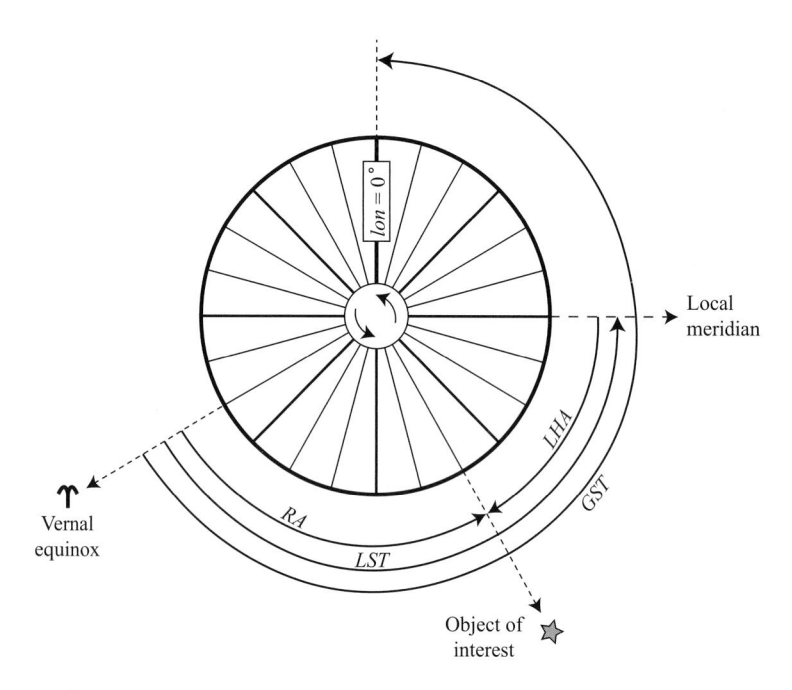

Figure 2.5. Illustration of *GST*, *LST*, *RA*, and *LHA*.
Credit: Michael K. Shepard.

being closer and then farther from the Earth as both orbited the Sun; the extra distance between observations added extra time to the apparent period.

When we observe any astronomical object, we are seeing it as it was in the past. The farther it is away, the farther back in time we are seeing it. Light from the Moon takes 1.25 s to reach the Earth, while light from the Sun takes around 8 minutes. This lag time can cause difficulties, and is especially acute for attempts to measure rotation periods or ephemerides.

For this reason, most ephemerides programs (like the Horizons program at JPL) provide the **light-corrected time** (**ltc**) of an event – the time *on Earth* when the light from the event left the object's surface. Conceptually, the time correction, Δt is based only on the relative distance between the observer (the Earth) and the observed (object), Δ, and can be calculated from

$$\Delta t = \frac{\Delta}{c} \qquad 2.13$$

where c is the speed of light. In practice, this correction is significantly more involved to account for relativistic effects. The light-corrected time is always *earlier* than the observation time.

2.3 Photometric Nomenclature

2.3.1 Terms and Symbols

The symbol and terminology conventions of photometry are not always consistent between the fields of physics and planetary science; indeed, even within the recent history of planetary science there have been inconsistencies. We will conform to the standards outlined in the USA Standard Nomenclature and Definitions for Illuminating Engineering, published jointly by the American National Standards Institute and the Illuminating Engineering Society of North America (IES, 2010). Where a particular term or convention is at odds with those used by the planetary community, this will be pointed out clearly with the most commonly used alternative symbol or term.

Table 2.1 lists the most commonly used photometric terms. From the perspective of a planetary scientist, the quantities most often used are **radiance** and **irradiance**, followed to a lesser extent by **radiant intensity**. Commonly, the reader will find that authors may use other terms and symbols for these same quantities. In many instances, a look at the units should be sufficient to determine which quantity is being referenced. However, some have similar units but different definitions. These will require a closer reading to make sure which is being discussed.

2.3.2 Irradiance and Flux Density

The **irradiance**, E, is the total amount of power, *over all wavelengths*, incident on a surface (e.g. a planet or a detector), divided by the surface area. It is therefore the light power per unit area and is in units of $W\ m^{-2}$. In most

Table 2.1 *Terms and Symbols*

Quantity	Symbol	Units	Definition
Flux Density	F	$W\ m^{-2}$	Power per area, perpendicular to travel
Irradiance	E	$W\ m^{-2}$	Power per unit area incident on surface
Spectral Irradiance	E_λ	$W\ m^{-2}\ nm^{-1}$	Irradiance per nanometer of wavelength
Radiant Intensity	I	$W\ sr^{-1}$	Power per solid angle
Radiance	L	$W\ m^{-2}\ sr^{-1}$	Power per unit surface area per steradian
Spectral Radiance	L_λ	$W\ m^{-2}\ nm^{-1}\ sr^{-1}$	Radiance per nanometer of wavelength

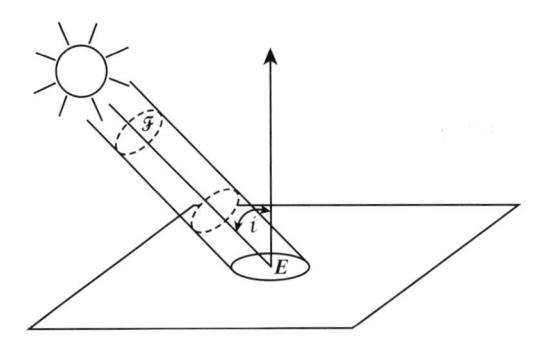

Figure 2.6. Illustration of the difference between flux density and irradiance. Credit: Michael K. Shepard.

instances, we are only interested in a given wavelength range – like those in the visible part of the EM spectrum – so we instead use the term **spectral irradiance**, which is given in W m^{-2} nm^{-1}, or other similar wavelength units[2]; in those cases, it is common practice to append a wavelength subscript to E if there is a chance for confusion.

The irradiance changes with the incidence angle because the power is spread out into an increasingly large area (Figure 2.6). To distinguish between the actual power per unit area incident on a surface (irradiance, E) and the true power per unit area perpendicular to the source, we use the term **flux density**[3] (F) for the latter. It has the same units as irradiance. The two quantities are related by the cosine of the incidence angle, a formula often referred to as Lambert's cosine law of illumination or the third law of photometry discussed in Chapter 1:

$$E = F \cos i = F\mu_0 \qquad 2.14$$

Historically, cos i is often abbreviated as μ_0, and the cosine of the emission angle (cos e) is abbreviated as μ.

It is commonly assumed that natural sunlight, the major source of illumination for objects in the solar system, is collimated. Because the Sun is not a point source, this is only an approximation, but it is a good one in most cases. At the Earth, the Sun subtends an angle of about 0.5°, so this is the deviation

---

[2] The units of wavelength vary widely. We have chosen to use nanometers, but one is just as likely to see micrometers, angstroms (10^{-10} m), or even meters.

[3] In planetary applications, it is common to see the flux density of the Sun written as πF instead of F for reasons to be discussed later.

from collimation. It can usually be neglected in all measurements except those very close to opposition (when the source and viewer are very close, phase angle ~0°). As we move out in the solar system, the approximation gets even better; at Jupiter, the Sun is only 0.1° in angular width.

2.3.3 Solid Angle

Before proceeding with the other terms, we must introduce or review the concept of **solid angle**, which has units of **steradians (*sr*)** (Figure 2.7). This is the three-dimensional equivalent of the angle in plane geometry. If we imagine every point source surrounded by a large spherical bubble, all the energy emitted must exit that bubble. In any given direction, we can envision a cone with its vertex at the center and the large end intersecting the bubble. By definition, the solid angle, Ω, subtended by the cone is the area, ***Area***, at the intersection with the spherical bubble divided by the square of the distance from the vertex, ***R***.

$$\Omega = \frac{Area}{R^2} \qquad 2.15$$

The surface area of a sphere is $4\pi R^2$, so it follows that the solid angle of the entire sphere is 4π *sr*, while that of a hemisphere is 2π *sr*.

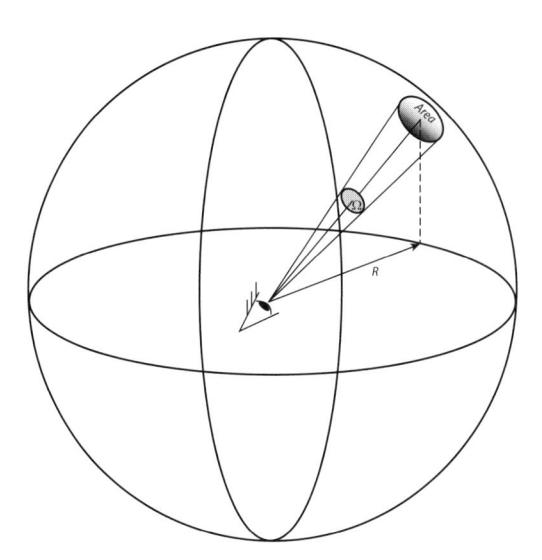

Figure 2.7. Illustration of the concept of solid angle.
Credit: Michael K. Shepard.

2.3.4 Radiant Intensity and Inverse Square Law

The quantity measured by a telescope or spacecraft in which the object is a point source, like an unresolved area or even an entire planet, is the **radiant intensity, *I*** (Figure 2.8). This is the power measured per unit solid angle of the receiver.

$$I = \frac{dP}{d\Omega} \qquad 2.16$$

Here we use the differential prefix, *d*, indicating a vanishingly small solid angle and power. This is because the measured power can vary with position on the sphere and with the size of the solid angle.

Going back to the idea of a sphere surrounding a point source of light, all of the power coming from the source is spread out evenly (assuming it's isotropic). As one moves outward, the area of that sphere expands with the square of the distance away from the source. The power, however, is a constant, so it must be spread more thinly. As a result, the power per unit area falls off with the inverse square of the distance from the source ($1/R^2$). This is commonly known as the **inverse square law** and was the second law of photometry discussed in Chapter 1. For a sensor with a fixed area, though, the perceived radiant intensity

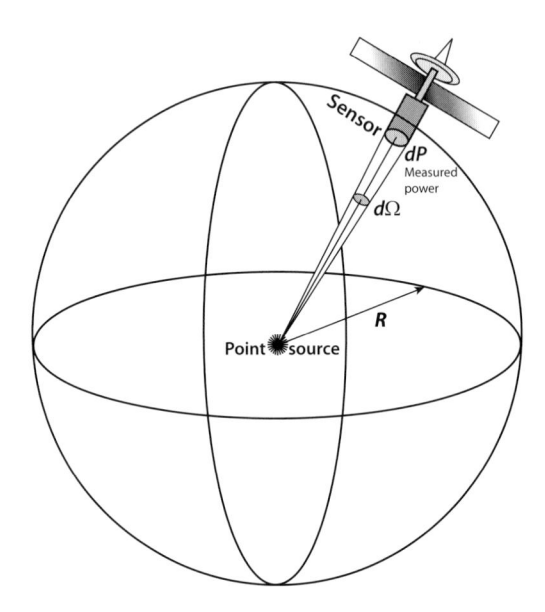

Figure 2.8. Illustration of the concept of radiant intensity, or more simply, intensity.
Credit: Michael K. Shepard.

will stay the same because the solid angle decreases at the same rate as the received power. So, the radiant intensity is a valuable quantity to measure because it is normalized, or independent of the distance from the source.

2.3.5 Radiance

The related quantity measured by a telescope or spacecraft in which the surface *can* be seen or resolved is the **radiance**,[4] L. Imagine a plane surface that reflects the light in all directions so that the incident light is reflected into an upper hemisphere of 2π steradians (Figure 2.9). A telescope or camera will only record the light exiting from a small part of that area through a small part of that hemisphere, so we normalize the measurement by area and solid angle, giving units of $W\ m^{-2}\ sr^{-1}$. We must include the effect of off-axis viewing because, if looking at a surface from the side at emission angle, e, the power reaching the sensor comes from a larger surface area than it would if looking straight down. As with the intensity, the measurement distance is irrelevant to the value of radiance. If farther away, the power per unit surface area intercepted by the telescope or sensor is smaller, but it is divided by a smaller solid angle.

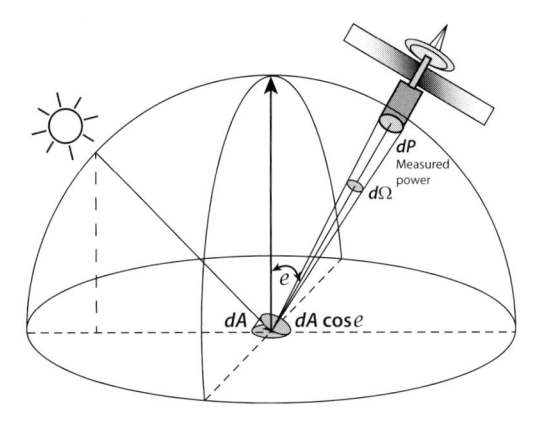

Figure 2.9. Illustration of the concept of radiance.
Credit: Michael K. Shepard.

---

[4] In many planetary references, radiance is often referred to with the symbol I, unfortunately confusing it with the standard symbol for radiant intensity. We will retain the standard radiance symbol L and point out when and where I is sometimes used in planetary literature.

The radiance at a point on the sphere is therefore the power measured at that point, dP, coming from a small surface area, dA, viewed at an emission angle, e, into a small solid angle, $d\Omega$, and is written as

$$L = \frac{d^2P}{dA\, d\Omega \cos e} = \frac{d^2P}{dA\, d\Omega\, \mu} \qquad 2.17$$

If the measurement is made in a specific wavelength interval, it is called the **spectral radiance** with units W m^{-2} sr^{-1} nm^{-1} or similar.

A relationship of some importance is that between radiance and radiant intensity. Given the definition of each, it is straightforward to show

$$I = L\,\mu\, dA \qquad 2.18$$

2.4 Magnitudes

2.4.1 Definitions

In Chapter 1, we introduced the magnitude scale based on the Pogson interval (Eq. 1.3). If two objects, A and B, vary by a factor of 2.512 in measured intensity, they differ by one magnitude. The relationship between the measured irradiance[5] (or radiant intensity) of any two objects and their magnitudes can thus be written

$$m_B - m_A = 2.5 \log_{10}\left(\frac{E_A}{E_B}\right) \qquad 2.19$$

or

$$\frac{E_A}{E_B} = 10^{-0.4(m_A - m_B)} \qquad 2.20$$

In planetary science, we distinguish between several kinds of magnitudes: **apparent, reduced,** and **absolute**. At the telescope, we measure the magnitude of a planetary object by comparing its measured irradiance to that of agreed upon standard stars and the application of Equations 2.19 and 2.20. This value is referred to as an **apparent magnitude** and will depend upon intrinsic properties of a planet, such as its size, intrinsic reflectivity, shape and orientation, as well as its distance from the Sun and from us (the observer), and the

---

[5] We will use the terms irradiance and radiance generically to include spectral irradiance and spectral radiance. We will be explicit in cases where there may be ambiguity.

2.4 Magnitudes

angles at which it is illuminated and viewed. Compared with stellar photometry, which simply standardizes stars to a common distance for comparisons, these factors are more troublesome to disentangle.

In an attempt to get at just the intrinsic properties of the planet, astronomers define the **absolute magnitude**[6] of an object to be its apparent magnitude if it were simultaneously 1 astronomical unit[7] (AU) from the Sun, 1 AU from Earth (or the observer if not on Earth), and fully illuminated. Note that this reference geometry is simply a convenient conceptual construct because it is physically impossible unless one places the observer deep inside the Sun! To derive the absolute magnitude from apparent magnitude requires an intermediate value – the **reduced magnitude**.

2.4.2 Reduced Magnitudes

The first step in the process of obtaining an absolute magnitude is to take a set of observations and *reduce* or transform them to the magnitude they would be if 1 AU from the Earth and Sun. This removes the effect of distance on a planet's brightness, although still leaving differences due to phase angle. The **reduced magnitude** $m(1, \alpha)$ is then defined as the magnitude of an object 1 AU from the Sun and 1 AU from Earth at phase angle α (note that we use only a single parameter to indicate the distance is 1 AU from both Earth and the Sun). The reduced magnitude is not the same as absolute magnitude because we do not specify that the phase angle be 0°. Instead we are only concerned with removing the effects of *distance* on brightness.

Let the apparent magnitude, m_{app}, of an object at a phase angle, α, a distance from the Sun, d, and distance from the Earth, Δ, be written explicitly as $m_{app} = m(d, \Delta, \alpha)$. We rewrite Equation 2.19 for a single object observed at arbitrary distances and 1 AU from the Earth and Sun as

$$m(d, \Delta, \alpha) - m(1, \alpha) = 2.5 \log \left(\frac{E(1, \alpha)}{E(d, \Delta, \alpha)} \right) \qquad 2.21$$

The amount of light the object receives from the Sun falls off as the inverse square of its distance from the Sun, d; likewise, its apparent brightness as

---

[6] In stellar astronomy, the same term is used. In that case, it means the magnitude a star would appear to be if it were placed 10 parsecs (32.6 light-years) from the Earth.
[7] The astronomical unit (abbreviations au and AU) was, at one time, defined to be the semi-major axis of the Earth's orbit about the Sun. In 2012, the International Astronomical Union, the governing body for astronomers around the world, divorced this unit from our actual orbit and defined it to be exactly 149,597,870,700 meters (nearly 150 million km) (IAU, 2012).

viewed from the Earth falls off as the inverse square distance from Earth, Δ, so that (if both d and Δ are in astronomical units):

$$E(d, \Delta, \alpha) = \frac{E(1, \alpha)}{d^2 \Delta^2} = \frac{E(1, \alpha)}{(d\Delta)^2} \qquad 2.22$$

Substituting this into Equation 2.21 and solving for $m(1, \alpha)$ gives

$$m(1, \alpha) = m_{app} - 5 \log d - 5 \log \Delta \qquad 2.23$$

If these are *visual* magnitudes (what would be seen by the human eye, Section 2.5), it is common to make this explicit using the notation V:

$$V(1, \alpha) = V - 5 \log d - 5 \log \Delta \qquad 2.24$$

Reduced visual magnitudes are usually plotted as a function of phase angle to generate a **phase curve**, a plot of fundamental importance for disk-integrated work (Chapter 6). A phase curve can be used to predict an object's magnitude at any other geometry. It can also be used to compare different objects in a standard way, and can reveal fundamental information about the physical properties of the object.

As an example, Figure 2.10 shows two plots of Venus's magnitude as a function of phase angle. The first plot shows the visual magnitude V (data from Mallama et al., 2006) without accounting for differences in distance. Note that it is brighter at phase angles >90° than at opposition, despite being only a crescent phase! The reason is the geometry. At high phase angles, Venus is closest to the Earth (~0.3 AU away), while at low phase angles, it is opposite the Sun (1.7 AU away). In the second plot, we have plotted the reduced magnitude, $V(1, \alpha)$, calculated using Equation 2.24. We will revisit this curve in Chapter 6.

2.4.3 Absolute Magnitudes

The **absolute magnitude** of a planet, written $m(0)$ (also sometimes $M(0)$), is the value we would observe if it were 1 AU from the Sun, 1 AU from the Earth, and at $\alpha = 0°$. The nomenclature here implicitly assumes unit distance from the Earth and Sun and explicitly shows $\alpha = 0°$. In the absence of any explicit wavelength identifier, it is assumed to be in the visual wavelengths and is often written as $V(0)$ and sometimes H or $H(0)$, especially for asteroids (Chapter 6). Unfortunately, we cannot directly measure $V(0)$ because we cannot meet all three observational conditions simultaneously. Although we can correct for distances other than 1 AU using Equation 2.24, *we can never observe an object at exactly $\alpha = 0°$*. Our own shadow would be in the way.

2.4 Magnitudes

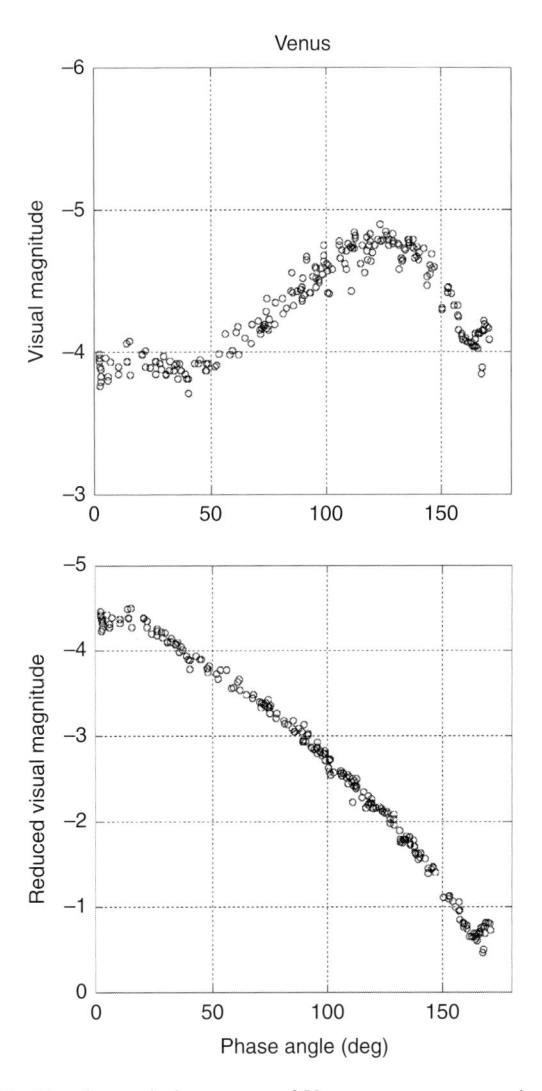

Figure 2.10. The observed phase curve of Venus – apparent magnitude versus phase angle. Once the apparent magnitudes have been corrected for reduced to a common distance (1 AU from the Sun and the Earth), the standard phase curve becomes evident.
Credit: Data from Mallama et al. (2006).

Every published value of absolute magnitude is therefore an *estimate* based on some method of extrapolation from observations near $\alpha = 0°$. For purposes of derivation, however, we will assume $m(0)$ (and other variants on this nomenclature) to *mean the true absolute magnitude* of an object.

For the superior planets, phase angles are limited to a fairly narrow range and it is common to linearly extrapolate to $\alpha = 0°$ and report this as the absolute magnitude (Hilton, 2005). For Mercury and Venus, nearly full phase curves (~0° – 180°) have been generated and fit with polynomial functions (Hilton, 2005; see Chapter 6).

If all objects were uniform spheres, this would be the end of it. But many objects – especially the smaller asteroids, comets, and some moons – are irregular in shape, and their brightness changes minute by minute as the object rotates. Fortunately, most objects spin about a single axis so that the amount of light they reflect to an observer varies in a regular, approximately sinusoidal way. The standard for absolute magnitude uses the mean brightness of an object as it makes one complete rotation. This normalization removes the variables of distance and rotation, but it doesn't completely remove the effects of shadowing on irregular shapes or the orientation of the object (its *spin axis*). We postpone discussing these complications until Chapter 6.

2.5 Photometric Systems

As we noted in Chapter 1, one of the major problems in early photometry was color. How can we compare two differently colored objects in brightness? In most of the work pre-1900, the problem was either ignored or addressed by attempting to filter one of the light sources until its color was close to the other. Lambert and others foreshadowed modern spectrophotometry by suggesting that we measure object brightness in individual colors – a **spectrum** (DiLaura, 2001). Those values could then be justifiably and accurately compared between different objects. Additionally, we can compare this spectrum of a star or planet with libraries of laboratory acquired spectra to determine something about composition.

Unfortunately, we are often forced to make a difficult choice between spectral resolution – the number of colors we measure – and the amount of light available for detection. This latter property is often referred to as the **signal-to-noise ratio** (*SNR*). The *noise* of a detector includes the unavoidable background signal from the source (e.g. sky brightness) and detector (e.g. thermal noise); the greater the signal compared to this, the more confident we are in the measurement. In measuring spectra, the finer the spectral resolution (the more colors), the fewer photons there are to measure. For bright objects, like the Sun, Moon, or even brighter planets, there is more than enough signal, even within a narrow range of wavelengths. This is usually not the case for

2.5 Photometric Systems

fainter objects, and is certainly not the case for early historical observations of color. The compromise is to use several broad-band filters, each in a different and relatively wide wavelength interval. This gives us some color information, and still allows in a significant amount of light so that signal-to-noise is reasonable.

A **photometric system** is a standardized set of well-characterized color filters and reference (standard) stars that allow investigators using different telescopes to directly compare their measurements. There are dozens of photometric systems, but here, we look at only a few of the more important ones for planetary photometry.

2.5.1 UBVRI Color System

The UBV photometric system (Johnson and Morgan, 1953), later expanded into the UBVRI system by **A.W.J. Cousins (1903–2001)** (Bessell, 1979), is the oldest and most venerable photometric system (Figure 2.11). It is often referred to as the **Johnson–Cousins** filter system and was developed from a desire to classify stars by temperature and color in a standard way so that astronomers working with different telescopes and photoelectric photometers or film could directly compare results.

Historically, the UBVRI band (short for "passband") filters were chosen for different reasons. The visual or V-band (center wavelength 550 nm) approximated the visual magnitude observed by eye. The blue B-band (center

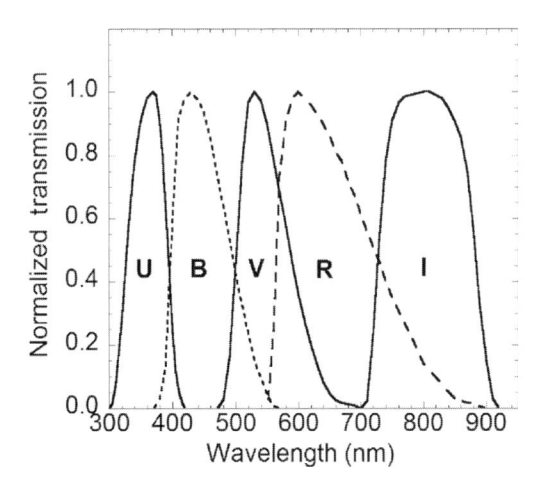

Figure 2.11. UBVRI passbands.
Credit: Data from Bessell (1990).

Table 2.2 *UBVRI Passbands*

Filter	λ_{cent} (nm)	$\Delta\lambda$ (nm)	m_{conv} (W m^{-2} nm^{-1})
U	367	66	−25.90
B	436	94	−25.36
V	545	85	−26.02
R	638	160	−26.89
I	797	149	−27.70

Filters, their center wavelength, passbands, and estimate of spectral irradiance (see Section 2.5.5) of stars reaching the Earth at the center wavelengths listed (Henden and Kaitchuck, 1990; Kitchin, 2003).

wavelength 440 nm) approximated the magnitude measured by the more blue-sensitive photographic plates (see Chapter 3), while the ultraviolet U-band (center wavelength 365 nm) was added because the hottest stars are brightest in the ultraviolet (UV), and a particular spectral feature, called the Balmer discontinuity, fell in the center of this band. The Earth's atmosphere (ozone) absorbs a significant fraction of the incident ultraviolet, so the short wavelength end of that filter is limited by the opacity of atmosphere. Unfortunately, this meant that the U-band brightness was affected by environmental processes that could not always be accounted for.

The red R-band (center wavelength 658 nm) and infrared I-band (center wavelength 806 nm) were later added to study cooler stars. Table 2.2 lists the filters, their peak wavelengths, and passband widths. Figure 2.11 illustrates their transmission characteristics (Bessell, 1990). Note that these have been normalized so that the peak transmission is 100%.

The standard for magnitude for the UBV system was originally defined by the A0 V (main-sequence) star Vega (α Lyra). By definition, Vega has **colors** or **color indices** of 0.0 in all of the above filters, where the word **color** or **color index** means the difference in magnitude between two filters. For example, the $(B - V)$ color index is $m_B - m_V$. For Vega, this is 0.0 by definition. Stars hotter than Vega emit more blue light (see Chapter 1, Section 3.1); this means m_B is smaller than m_V (remember, smaller magnitudes mean brighter), so their color index is negative; for cooler stars, it is positive. For reference, the Sun has a $(B - V)$ index of 0.656 ± 0.005 (Gray, 1992). Originally, Vega was defined to have a visual magnitude of 0.0, but today it is agreed that it has a magnitude of $m_v = 0.03$ (Bessell, 2005). We look at modern standard stars in Section 2.6.

2.5 Photometric Systems

Table 2.3 *JHK Passbands*

Filter	λ_{eff} (nm)	λ_{eff} (nm)	2MASS λ_{eff} (nm)	2MASS $\Delta\lambda$ (nm)
J	1220	213	1235	162
H	1630	307	1662	251
K	2190	390	2159 (K_s)	262

Bandwidth centers and full-width half-maxima are shown (Kitchin, 2003).

2.5.2 Infrared JHK System and Two Micron All Sky (2MASS) Survey

The advent of infrared telescopes required an extension of the UBRVI system deeper into the infrared, which became the Johnson/Cousins JHK (and later extended to LMN) system (Bessell, 2005). The band centers and widths are often, but not always, constrained by atmospheric absorption due to the presence of water vapor. Each infrared observatory has their own, slightly different version of each filter that must be transformed into a standard JHK magnitude.

The 2MASS survey (1997–2001) was an all-sky survey using 1.3 m telescopes with the near-infrared J, H, and K bands (Skrutskie et al., 2006). Telescopes were sited in Arizona and Chile to cover both the northern and southern hemispheres, and the survey was designed to answer a number of astronomical questions. The survey swept up thousands of asteroids and comets, and the 2MASS colors aid in their classification (Sykes et al., 2000, 2010). The standard JHK and 2MASS JHKs filter effective centers and passbands are listed in Table 2.3.

2.5.3 Gunn "griz" System and Sloan Digital Sky Survey

While the UBVRI, etc., is probably the most popular photometric system because of its heritage, there are others. The Gunn system is based on a set of filters – u, g, r, i, and z – and standard stars originally defined in the late 1970s at Hale Observatories in California (Thuan and Gunn, 1976; Wade et al., 1979). The goal of this project was to define non-overlapping filters that avoided the brightest light pollution emission lines and matched the properties of films used at the time. For standards, they wanted stars with flat spectra and few absorption features in the visible and settled on slightly cooler F-class stars instead of the hot A0 stars. The 200 inch (5 m) Hale telescope routinely used the F-class star BD +17° 4708 as a principal standard for its

Table 2.4 *Original SDSS Passbands (Gunn et al., 1998)*

Filter	λ_{eff} (nm)	$\Delta\lambda$ (nm) (FWHM)
u′	354.9	56.0
g′	477.4	137.7
r′	623.1	137.1
i′	761.5	151.0
z′	913.2	94.0

spectrophotometer, so this star was adopted and assigned a magnitude of $g = 9.500$ and all its colors (g–r, etc.) were defined to be 0.0.

This system was adopted and slightly modified by the Sloan Digital Sky Survey (SDSS) (Gunn et al., 1998). This survey consists of a 2.5 m wide-field telescope to survey much of the visible northern sky down to ~23rd magnitude, multiple times, in multiple wavelengths, primarily for stellar, galactic, and cosmologic investigations. However, the survey also sweeps up thousands of asteroids and Kuiper belt objects and is often utilized by planetary scientists because the SDSS colors can be used to differentiate between some asteroid classes (e.g. Ivezic et al., 2001). Like the UBVRI system, the Sloan survey uses five filters covering the same wavelength regions with slightly different emphases. Table 2.4 lists their effective band centers and widths. There are caveats in the use of this system because two slightly different filter sets were used, and the reader is encouraged to consult Bessell (2005) and references within for explanations.

2.5.4 Comet IHW and HB Narrowband Filter Systems

Because of their icy nature, comets require a different set of photometric systems. As they approach the Sun, the ice in the nucleus vaporizes and photodissociates, creating a fuzzy **coma** full of numerous daughter molecules that fluoresce and emit narrow spectral bands on top of the underlying reflectance spectrum, or **continuum**, of the nucleus and entrained silicate components. Measurement of these individual **emission bands** provides information on the type and abundance of the components making up the nucleus, and the production rate of gas and dust (Schleicher and Farnham, 2004).

In the early 1980s the IAU Commission 15 Working Group recommended the development of a nine-filter set, specifically designed to capture the six most important emission bands known at the time (C_2, C_3, CN, CO^+, H_2O^+, OH)

2.5 Photometric Systems

and three continuum points (Osborn et al., 1990). The work was important because of a worldwide effort to observe and characterize Comet Halley during its 1986 apparition, so this particular set of filters is referred to as the International Halley Watch, or IHW filters.

Degradation of the original filters over the next decade and the discovery that there were contamination issues – where the tails of an unintended emission band bleed over into the window of a particular filter – led astronomers to develop an improved set of filters in time for observations of Comet Hale–Bopp in 1995. This newer set is referred to as the HB (Hale–Bopp) filters. It consists of 11 filters that largely overlap the earlier IHW filters, plus a new filter for the NH emission line and one additional continuum band (Farnham et al., 2000).

2.5.5 Absolute Flux Units and the AB Magnitude System

So far, we have defined magnitude and color only in terms of specific filter sets and standard stars that everyone agrees are such-and-such magnitude or have a certain color index. We measure by comparison. But what is the actual relationship between a measured irradiance and the magnitude of a star? Absolute calibration of this sort is used by planetary scientists in the radio and microwave spectrum, but not commonly in the **visible–near infrared** (VISIR, wavelengths of ~400–2500 nm or so). Nevertheless, it is useful to be aware of it.

Table 2.2 has a column labeled m_{conv}. This is reproduced from Henden and Kaitchuck (1990) who in turn obtained it from Johnson (1965). For any given magnitude at a particular wavelength, m_λ, we find the spectral irradiance from

$$E_\lambda = 10^{-0.4(m_\lambda - m_{conv})} \qquad 2.25$$

The values in Table 2.2 are conversion factors for wavelengths near the center of their respective filters. The total irradiance on a sensor can be approximated by

$$E \approx \Delta\lambda \, E_\lambda \qquad 2.26$$

where $\Delta\lambda$ is the width of the filter (in nm) and we assume it is not too wide compared with the distance between filter centers.

As an example, the star Vega with a visible magnitude of 0.03 would have a spectral irradiance of

$$E_\lambda = 10^{-0.4(0.03 + 26.02)} = 3.80 \times 10^{-11} \, W \, m^{-2} \, nm^{-1} \qquad 2.27$$

in the visual range, and if measured with that filter (assuming 100% bandpass efficiencies at all wavelengths within the filter), we would expect a total irradiance of

$$85 \text{ nm} \times 3.80 \times 10^{-11} \text{ W m}^{-2} \text{ nm}^{-1} = 3.23 \times 10^{-9} \text{ W m}^{-2}$$

The **AB magnitude system**, unlike the others discussed, is based upon absolute flux measurements like these, and the zero point is defined to be 3,631 Jansky (Jy), where the Jansky is a commonly used measurement of flux (especially among radio astronomers) and 1 Jy = 10^{-26} W Hz^{-1} m^{-2} (Bessell, 2005). Although now divorced from the flux of any particular standard star, the system was originally defined based upon a mean calibration and flux model of Vega (Oke and Gunn, 1983). Conversions between the AB system and others are widely available in the online documentation of major observatories.

In all of these matters of photometric systems, we have glossed over the very real technical issues associated with making precise measurements. Most telescopes look through the Earth's atmosphere, and its thickness and the associated diminution of light depends upon the angle above the horizon (greater thickness of atmosphere near the horizon, lesser near the zenith), the altitude, the weather, and other factors. And every telescope and filter is different from others in use, and every sensor has a slightly different response to incident light, or **response function**. As a result, every set of observations must go through a rigorous set of corrections for these factors to allow direct intercomparison of measured fluxes and magnitudes between different observatories. This is done regressively by using a wide variety of standard stars. A good review of these methods can be found in Henden and Kaitchuck (1990) and Warner (2006). Most specialists in this field will tell you that, despite years of work, it is still difficult to get photometric results accurate to better than a few percent. We revisit this in more detail in Chapter 6.

2.6 Standard Stars, Solar Analogs, and the Zero Point Magnitude

In many applications, we must reference our observations to **standard stars**. In the earliest stellar surveys, a single star, e.g. Polaris, was often used as the standard and defined to have a specific magnitude, and sometimes color. In practice, this system has two limiting conditions: the standard star must be visible whenever a comparison must be made for best results, and you may wish to compare it to an object that differs significantly in color. The first condition restricts observers to using one or more stars in the circumpolar region of their respective hemisphere. The second limits observers to comparison with similarly colored objects.

2.6.1 Early Standards and the Landolt Stars

In the nineteenth century, the Potsdam photometric survey chose the more complex but versatile method of using a variety of standard stars of different colors scattered about the sky, all between magnitudes 4.5 and 7.3 (Hearnshaw, 1996). In his original survey, Pickering had considered using ten stars near Polaris in much this same way. Later, he would realize this concept in the **North Polar Sequence**, a group of some 100 stars near Polaris with a wide range of magnitudes and colors. The work on this sequence was primarily the responsibility of **Henrietta Leavitt** (1868–1921) who is perhaps more famous for working out the relationship between the periods of Cepheid variable stars and their intrinsic brightness (Hearnshaw, 1996).

With the advent of the modern UBVRI photometric system, it became important for a number of well-characterized stars to be identified as standards, both in magnitude and color. Today, the principal set of standard stars in use were chosen and characterized by **Arlo Landolt** (Louisiana State University) (Landolt, 1983, 1992; Landolt and Uomoto, 2007a, b). On the order of a thousand stars have been characterized and several hundred are routinely used as standards in astronomical work. The SDSS survey also has a set of standards, many of which overlap the Landolt list (e.g. Smith et al., 2002).

2.6.2 Solar Analog Stars

Planetary photometry differs from stellar photometry in that we are looking at sunlight reflected from an object. For many applications, like generating lightcurves or determining phase angle behavior, nearly any standard star will suffice. However, if the goal is to determine the reflectance properties of the object, especially in multiple wavelengths (**spectrophotometry**), it is important that the standard star be as much like the Sun as possible; otherwise, the reflectance we measure could be seriously in error. Unfortunately, the brightness of the Sun precludes us from measuring it directly as a standard. Instead, a significant amount of work has gone into finding stars that have very similar temperature, compositional, and spectral characteristics. Much of this work began with **Johannes Hardorp** (Hardorp, 1978, 1980) and continues to this day (Hall, 1997; Onehag et al., 2011).

A **solar-like** or **solar-type** star is one which shares many of the Sun's characteristics, including colors and temperatures. Soderblom and King (1998) note that a wide range of stellar types, from F8 to K2 fit these criteria. A more restrictive category is the **solar analog** star, defined to be one with no close companion stars, a temperature within 500 K of the Sun, and a **metallicity**, defined as the abundance of everything except H and He, within a factor of two of the Sun (Soderblom and King, 1998). A rarer class of star is the **solar twin**, defined as one that has no companion stars, temperatures within 10 K of the Sun, metallicity within ~10% of the Sun, and of similar age to the Sun (Soderblom and King, 1998). For most planetary spectrophotometric work, solar-like stars are sufficiently close in properties to the Sun to be used as reflectance standards.

For comet investigations, an additional set of standard stars are needed – **flux standards**. These are used to determine the amount of atmospheric extinction present and provide for a conversion from relative magnitude to absolute flux (Farnham et al., 2000), useful for estimating gas emission abundances. The chief criteria for these stars are that they be stable, bright in the UV (where many comet emission lines are found) and have few spectral features. This largely limits them to the hotter O- and B-type stars (Farnham et al., 2000). Ideally, a large number of these widely spread about the sky can be identified to make it relatively easy to find one at an air mass similar to that of the comet under investigation.

2.6.3 A Magnitude Standard

Up to this point, we have glossed over a troubling fact in photometry: until very recently, there was no internationally agreed upon standard for magnitude. As we noted earlier, the A0 star Vega is nominally the standard for the UBVRI system with $m_v = 0.03$ and colors defined to be 0.0. And the Gunn/griz/SDSS system uses the F-class star BD +17° 4708 as a principal standard for magnitude and colors. But these are conventions, agreed upon informally and propagated by continued use – not formally defined standards. This changed in August 2015 when the International Astronomical Union (IAU) adopted a zero point standard for magnitude at their 29th General Assembly (IAU, 2015). Before describing the new definition, we must briefly describe the magnitude scale used in the definition – the **bolometric magnitude** scale.

The bolometric magnitude scale is based on the total energy emitted by a star, over all wavelengths. It is often assumed that this energy can be computed by assuming the star to behave as a blackbody at some effective

2.6 Standard Stars, Solar Analogs, Zero Point Magnitude

temperature (see Chapter 1). So the bolometric magnitude is not one we can see, but one we would measure if we could account for all the energy emitted by a star. For solar-like stars, the bolometric magnitude and visual magnitude are similar because solar-like stars emit most of their energy in the visual wavelengths. For hotter and cooler stars, however, the visual and bolometric magnitudes may differ considerably and conversion between the two requires a correction factor, k

$$m_v = m_{bol} - k \qquad 2.28$$

Written this way, the correction factor k is always a negative number, meaning the visual magnitude is always larger (dimmer) than the bolometric magnitude. For stars like the Sun, $k = -0.08$, so the visual and bolometric magnitudes are nearly the same. But for stars like O-supergiants, $k \sim -4.0$ because most of their energy is in the ultraviolet (Kaler, 1997).

IAU Resolution B2 recommended that the zero point of the **absolute bolometric magnitude** scale, $M_{bol} = 0$, be defined by a source with a luminosity of 3.0128×10^{28} W. You may recall (see footnote in Section 2.4) that for stars, absolute magnitude means the magnitude we would observe if the star in question were 10 parsecs (32.6 light-years) from Earth. Therefore, a star with $M_{bol} = 0$ would have an irradiance $E = 2.518 \times 10^{-8}$ W m^{-2} at the Earth. Why was this odd number chosen as a definition?

The Sun has been the subject of innumerable measurements by solar astronomers for more than a century. And while it is not considered a variable star, its luminosity does fluctuate at the 0.1% level (Wilson and Hudson, 1991). Its average (or nominal) luminosity is 3.828×10^{26} W, and at a distance of 1 AU from it and just above the Earth's atmosphere, we receive, on average, an irradiance of $F = 1,361$ W m^{-2}. If our Sun were moved to the standard 10 parsecs from the Earth, it would have an absolute bolometric magnitude of $M_{bol} = 4.74$. The new IAU definition for the zero point was chosen, at least in part, because it is consistent with these average values for the Sun. Now, *by definition*, the Sun has an absolute bolometric magnitude of $M_{bol} = 4.74$, a flux density $F = 1,361$ W m^{-2}, and an apparent bolometric magnitude of -26.832. If you apply a bolometric correction factor of $k = -0.08$ to get visual magnitudes for the Sun, we find its absolute magnitude to be $M_v = 4.82$ and apparent visual magnitude to be $m_v = -26.75$, both values that are widely reported in the literature (Cox, 2000).

This type of definition resembles the IAU's earlier decision to divorce the definition of the astronomical unit from an actual property of nature – in that case, the semi-major axis of the Earth's orbit. There, as with this definition, a value was chosen that was both exact and well approximated the quantity in question to within experimental uncertainties.

3
The Mechanics of Planetary Observing

Light from a distant planet has a perilous journey before reaching our detectors. It must first leave the Sun, travel millions of kilometers to the object of interest, scatter from its atmosphere or surface, escape the object without additional scattering or absorption, travel along a very precise trajectory toward the Earth, and survive its transit through our atmosphere before finally being collected by an optical device and recorded. Uncountable numbers of photons leave the surface of the Sun every second, and if lucky, we collect a few thousand reflected from our target.

In this chapter, we discuss the mechanics of observing, beginning with the most common telescope/camera configurations in use for ground-based and spacecraft observations. We then look at the detection and recording systems used (spending the bulk of the time on CCDs), and finish with an overview of how the raw data from these tools are eventually turned into photometric quantities.

3.1 Telescopes

In this section, we look at the optical systems – telescopes and cameras – that do the observing. We begin with a brief overview of the common mounting systems. Although the mounting and pointing systems for space-based telescopes are beyond the scope of this book, the common ground-based mounting methods are worth a review. The remainder of the section describes common optical configurations for telescopes and the relevant properties needed for photometric work.

3.1.1 Telescope Mounting

Ground-based telescopes must be able to point to and track an object of interest, and there are two standard methods to mount them and numerous variations on these themes. The simplest mounting method is the **alt-azimuth mount**. In this design, a telescope has two axes of motion: rotation about the vertical axis (side-to-side or left-to-right) and about the horizontal axis (up and down). This is how the earliest telescopes were mounted. The main problem with this design is that the axes of motion do not match the rotation of the sky about the celestial pole. To track an object for long-term observing or imaging, both telescope axes must be moved at varying rates to maintain pointing. Even if this is done correctly, the image-plane itself rotates, and a fixed camera would see the image rotate.

To address the difficulties of tracking with the alt-azimuth mount, most observatories use an **equatorial mount**. A simple way to think about this mounting design is to tilt the alt-azimuth mount so that its (formerly) vertical axis points to the north or south celestial pole (depending on the hemisphere it is in). This axis, now referred to as the **polar** or **declination** axis, is aligned with the spin axis of the Earth, and one only needs to rotate the telescope along an east-west path to track an object. Since the Earth's spin rate is constant, it is a simple matter to put a drive motor on this axis to exactly track a target. This design also removes the problem of the rotating image-plane (Figure 3.1).

Large telescopes on equatorial mounts have significant engineering challenges because of their asymmetric mass distributions, but until recently, these challenges were preferred over the greater ones presented by the alt-azimuth mount. However, advances in computer technology have largely eliminated the advantages of the equatorial mount. The largest modern telescopes, like the 10m Keck telescopes, sit on alt-azimuth mounts because they are structurally more efficient. Computers easily drive motors on both rotation axes at the proper rates, and there is an additional computer-controlled motor rotating any devices at the image-plane to maintain accurate alignment.

3.1.2 Telescope Types

3.1.2.1 Refractors

The earliest optical systems used by both telescopes and cameras were based on the refraction of light through optically clear glass and are called **refractors** (Figures 3.1 and 3.2). Unfortunately, refraction through glass varies with the wavelength of light, so the images produced in early refractors suffered from **chromatic aberration**, evident as a colorful halo around a bright star or planet. This was largely alleviated by the invention of achromatic lenses by **Chester Hall** (1703–1771) around 1730. But there are other

Figure 3.1. The 36-inch (0.9 m) James Lick refracting telescope on equatorial mount. Public Domain. www.loc.gov/pictures/item/2013645277/

challenges with refractors, and an alternative design, the **reflecting telescope**, was invented by Isaac Newton in 1668 (King, 1955). In this design, light is gathered and focused by a concave mirror (Figure 3.2).

Reflecting telescopes have many advantages over refractors. The optical substrate does not have to be perfect or even transparent; all of the optical activity occurs at the mirrored surface. Although they have their own image quality issues, reflectors do not suffer from chromatic aberration. They have far fewer optical surfaces to grind and polish. And perhaps most importantly, they have fewer engineering issues and can be much larger than refractors. The largest operational astronomical refractor in the world, Yerkes Observatory, has an objective diameter of 40 inches (1.0 m), while the largest reflecting telescopes are currently the Keck 10m giants on Mauna Kea. For these reasons, all major observatories and most modern spacecraft cameras employ some

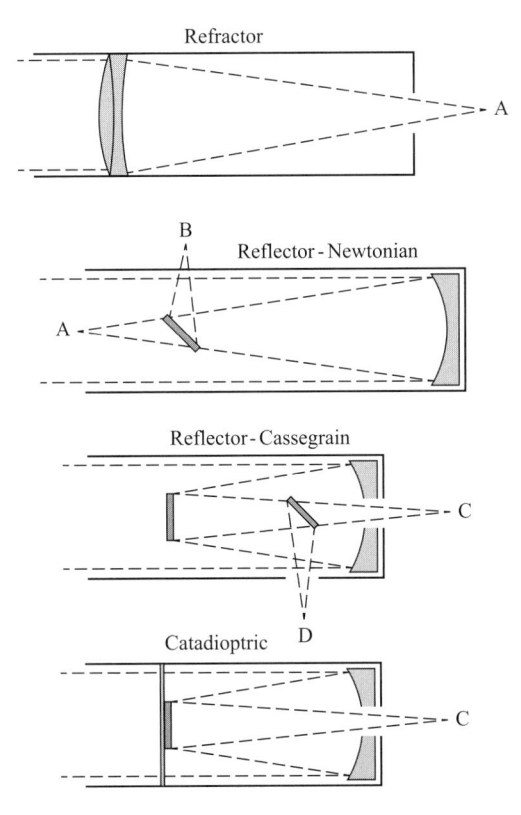

Figure 3.2. Schematic of different reflecting telescope configurations. Points labeled A are at the prime focus, B is the Newtonian focus, C points are at the Cassegrain focus, and D is at the Nasmyth or coudé focus.
Credit: Michael K. Shepard.

form of reflecting telescope architecture. A recent notable exception to this are the Dawn Framing Cameras, which each use a multi-lens refraction system (Sierks et al., 2012).

3.1.2.2 Reflecting Telescopes

Newton's early telescope became the prototype for an entire class referred to as the **Newtonian telescope**. In its earliest form, a **primary mirror** was ground to have a concave spherical shape to focus light to a point. Light reflected from the mirror leads to a focal point some distance, called the **focal length** (*fl*), away. With compound telescopes (e.g. Cassegrain, see below) the focal length of the primary mirror is often increased or decreased with auxiliary optics and the **effective focal length** (*efl*) is the property of interest. The ratio of the effective focal length to the primary or **objective diameter**, *D*, is called the focal ratio, often abbreviated *f/r*:

$$f/r = \frac{efl}{D} \qquad 3.1$$

An *f*/5 telescope means the effective focal length is five times the diameter of the objective.

Virtually every reflecting telescope design places some object, mirror or instrument, within the path of the incoming light (Sidgwick, 1971). In a Newtonian, a small **secondary mirror** is placed in the path of the reflected light and diverts it to the side so an observer or larger instrument can be placed there without blocking the incoming light. In many larger observatory telescopes, the instrument or camera is similar in size to the secondary mirror, and some provide for placing these at the **prime focus**, within the telescope itself (Figures 3.2 and 3.3).

The placement of an object within the optical path of the mirror creates two minor problems: (1) it diminishes the total amount of light available, and (2) it creates a diffraction effect around stellar images. The first problem is inevitable unless one uses a tilted mirror arrangement. This creates its own issues, but has been pursued by advanced amateur telescope makers and was used in the MESSENGER MDIS Narrow Angle Camera (Hawkins et al., 2007). The second issue is worse for bright objects and often shows up, even in the best observatory images, as a series of thin **diffraction spikes** that radiate from any bright source.

The first Newtonian telescopes used concave spherical mirrors. Unfortunately, even a perfect mirror of this type will not focus collimated light to a point, a defect referred to as **spherical aberration**. This problem is minimized in very long focal ratio telescopes, but is exacerbated as the focal ratio is reduced. It can be fixed by slightly changing the figure of the mirror from spherical to parabolic, and all modern Newtonians take this approach. Unfortunately, this causes a new problem called **comatic aberration**, or **coma** for short. While the parabolic mirror gives excellent stellar images along the optical axis, they become smeared off axis and appear to develop a tail (like a comet). The further off axis, the greater the coma. Despite these issues, the Newtonian was the observatory instrument of choice for centuries and is still a favored design for many amateur astronomers (Sidgwick, 1971).

The **Cassegrain** reflector, the design of which is attributed to **Laurent Cassegrain** (~1629–1693) in 1672 (Baranne and Launay, 1997), replaces the flat secondary mirror with a hyperbolic convex mirror that diverts the reflected

3.1 Telescopes 69

Figure 3.3. Catalina reflecting telescope on an equatorial mount. Camera at prime focus. Cassegrain focus is covered.
Credit: Catalina Sky Survey, Lunar and Planetary Laboratory, University of Arizona. Used with permission.

light back through a central hole in the primary mirror. This general type of design has several advantages, among them a greater ease of mounting and balancing heavy instruments at the focus, and a longer focal length in a more compact telescope.

One variant of the Cassegrain is called the **Nasmyth**, after **James Nasmyth** (1808–1890), who placed a third flat mirror in the path of the secondary reflection to direct the light to the side, as in the Newtonian, but farther down the main structure. A variant of this design, called the **coudé** ("elbow" in French) aligns the optical path of the third reflecting surface with the declination axis (in an equatorial mount) of the telescope. This allows one to mount a stationary instrument on the observatory floor, which will not need to be supported or moved by the telescope drive system.

3.1.2.3 More Complex Designs

A variety of smaller telescopes, like those used by many amateur astronomers, employ combinations of correcting lenses with mirrors and are generically referred to as **catadioptric telescopes**. A favorite design is the **Schmidt–Cassegrain**, partially named for **Bernhard Schmidt** (1879–1935), a legendary telescope maker and inventor of the wide-field Schmidt camera. In the Schmidt–Cassegrain, the primary and secondary mirrors are (often) spherical – making them inexpensive to mass produce – and the secondary is mounted in the center of a clear **Schmidt corrector plate**, a thin lens that corrects for spherical aberration. Other popular designs include the **Maksutov–Cassegrain** (after **Dmitry Maksutov**, 1896–1964), which employs a large negative meniscus lens instead of a Schmidt corrector plate. One major advantage of these systems is that the corrector plate allows the telescope to be sealed at the front and rear, making for a very rugged design.

No optical system is perfect, but with more complex mirror figures, problems can be reduced. The optical system most commonly used in modern observatories and spacecraft cameras is the **Ritchey–Chrétien** (RC) telescope, invented by astronomers **George Ritchey** (1864–1945) and **Henri Chrétien** (1879–1956). It is a variant of the Cassegrain and employs a concave hyperbolically figured primary and hyperbolic convex secondary. It is known for having a relatively wide field free of coma and spherical aberration. Examples of its use include the Keck 10m telescopes, the LORRI imaging camera on New Horizons, and the LROC, the imaging camera on the Lunar Reconnaissance Orbiter (Keck, 2002; Cheng et al., 2008; Robinson et al., 2010).

Perhaps the most advanced optical system in current use by astronomical telescopes or spacecraft is the **three-mirror anastigmat**. In this design, three curved mirrors are used to produce a wide field of view while reducing as many optical aberrations as possible. Examples of this in use are the HiRISE camera on the Mars Reconnaissance Orbiter (McEwen et al., 2007), the James Webb Space Telescope (Nella et al., 2004), and the Large Synoptic Survey Telescope (LSST; Olivier et al., 2008).

3.2 Optical System Properties

All optical systems can be characterized by a few basic quantities: objective diameter, D; effective focal length, efl; focal ratio, f/r; resolution, θ; field of view, FOV; and limiting magnitude, m. The diameter, focal length, and focal ratio are strictly a function of the telescope or camera optics. All of the others are functions of both the optics *and* the detection system.

3.2 Optical System Properties

3.2.1 Field of View

The field of view (*FOV*) of an optical system is an angle corresponding to the apparent width of the image, and can be found if one knows the focal length and the physical size of the detector. Figure 3.4 shows the geometry, and inspection shows that the *FOV* can be found from

$$FOV = 2\tan^{-1}\left(\frac{w/2}{efl}\right) \quad 3.2$$

where *w* is the width of the detector along one dimension. If the detector is rectangular, the *FOV* will differ along the two axes. If *efl* » *w*, then a good approximation to the above equation is

$$FOV = \frac{w}{efl} \quad 3.3$$

where *FOV* is in radians. As an example, consider a 35 mm camera with a 50 mm objective with 300 mm focal length. The focal ratio is 300/50 = *f*/6. The 35 mm film has active dimensions of 24 mm × 36 mm. The *FOV* of such a camera is 4.6° × 6.9°, or 80 × 120 milliradians (mrad).

3.2.2 Point Spread Function and Resolution

When collimated light is focused by a telescope, it does not converge to an infinitesimal point. Even if the optical system is perfect, diffraction effects

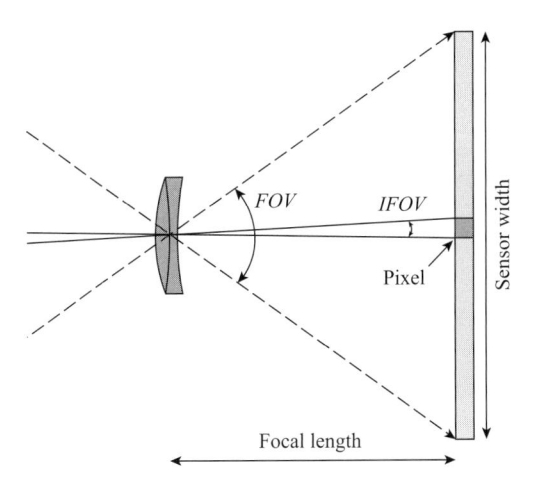

Figure 3.4. Schematic of the geometry of field of view and instantaneous field of view.
Credit: Michael K. Shepard.

from a circular aperture lead to a symmetric series of concentric rings called the **Airy pattern**, named for **George B. Airy** (1801–1892) (Figure 3.5; Airy, 1835). The center of these rings is a spot, called the **Airy disk**, and a three-dimensional plot of the intensity of this pattern (an Airy function) is cone-like and can be closely approximated by a Gaussian function (Figure 3.6).

The **point-spread function** (*PSF*) of an optical system is the mathematical representation of the output at the telescope focal plane if the input is a perfect point. For most high-quality telescopes and cameras, the optics are such that the *PSF* can be well represented by a Gaussian function. A common parameter for describing the *PSF* is its **full-width at half-maximum** intensity, or

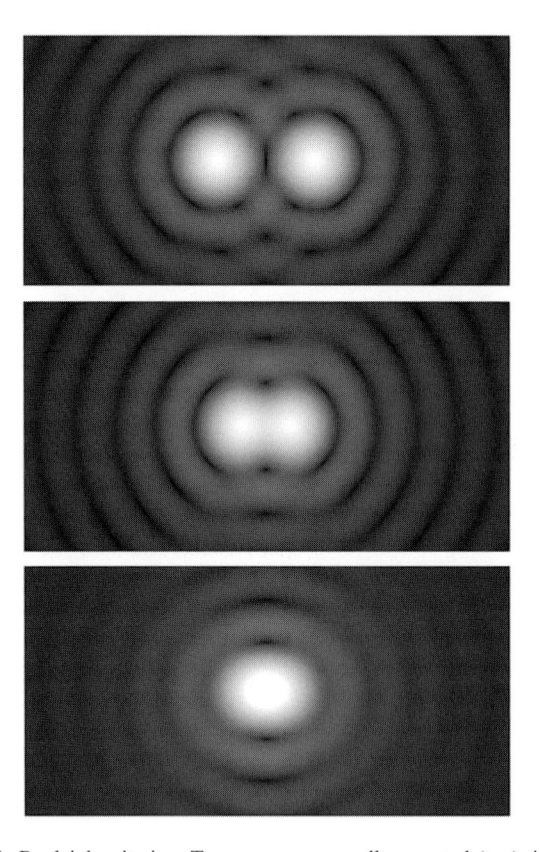

Figure 3.5. Rayleigh criterion. Two sources are well separated (top), just separable (middle), or indistinguishable (bottom). The just separable points meet the Rayleigh criterion.
Credit: Spencer Bliven, Public Domain, https://commons.wikimedia.org/w/index.php?curid=31456019

3.2 Optical System Properties

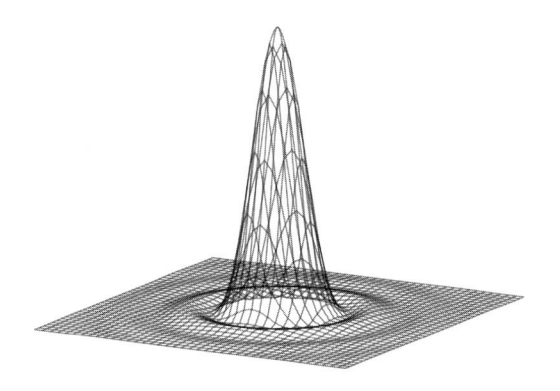

Figure 3.6. Three-dimensional plot of a typical Airy function. Its true form is a Bessel function, but it can be well represented by a Gaussian function.
Credit: Sakurambo, Public domain. https://en.wikipedia.org/wiki/File:Airy-3d.svg

FWHM. If the *PSF* can be approximated by a Gaussian function, the *FWHM* is ~2.4 standard deviations wide (2.4σ, or $\pm 1.2\sigma$ from the mean).

The resolution of an optical system is determined by its *PSF*. The Airy disk (up to the first minimum in the Airy pattern) in a perfect optical system has an angular width

$$\theta \cong 1.22 \frac{\lambda}{D} \qquad 3.4$$

where θ is in radians, λ is the wavelength, and $D \gg \lambda$. In the event two point sources are separated by the angular width of the Airy disk, they will be seen as different objects. This angular width is considered to be the **limiting resolution** of an optical system. In many references for optical telescopes this limit (and slightly different versions of Eq. 3.4) may be referred to as the **Dawes' limit** (for William R. Dawes, 1799–1868) or the **Rayleigh criterion** (for John William Strut, aka Lord Rayleigh, 1842–1919) (Figure 3.5).

Equation 3.4 gives the resolution in radians, but the values are so small for most optical systems that it is more common to see them expressed in **milliradians** (mrad) and **microradians** (μrad). Resolution is also commonly given in terms of **arcminutes** (′) or **arcseconds** (″), where there are 60 arcminutes (′) in a degree (°), 60 arcseconds in an arcminute, and 3,600 arcseconds in a degree. The resolution of the human eye is on the order of 1′, and a small telescope of 50 mm aperture is ~2.3″. For conversion purposes

1 mrad = 3.44′ (206.3″) or 1′ = 0.29 mrad
1 μrad = 0.21″ or 1″ = 4.85 μrad

Just because a telescope has a specific limiting resolution does not mean that it can or will be achieved, even assuming the optics of the telescope or camera are near perfect. On Earth, resolution is usually limited by instabilities in the atmosphere, a factor referred to as **seeing**. Seeing can be defined as the angular diameter of a star when perfectly focused in a large telescope, or more technically, the *FWHM* of the point spread function. In practice, it is also the angular resolution limit at that time and place, regardless of the telescope type. For example, a typical suburban or rural site may have a seeing limit of 2″. This means that, no matter how big the telescope, the best resolution attainable at that time is 2″.

On mountains at the best observatories in the world, the seeing is rarely better than 0.4″ which is the limiting resolution for a 0.3 m telescope. To approach the limiting resolution of a major telescope, one must either put it in space (like the Hubble Space Telescope) or utilize adaptive optic technology to minimize the atmospheric problem. And with a modern CCD detector (discussed later), the pixel size plays a very important role in achieving the best resolution possible.

3.2.3 Limiting Magnitude

The limiting magnitude of a telescope system is dependent upon the sensor and the exposure time. Assuming that one knows the limiting magnitude of a reference sensor, m_0, with an aperture of D_0, the limiting magnitude of another sensor, m_1, of aperture D_1 can be estimated (assuming all other conditions are identical) by comparing the light gathering *areas*:

$$m_1 = m_0 + 5 \log \left(\frac{D_1}{D_0} \right) \quad \quad 3.5$$

Prior to the advent of artificial sensors, e.g. photographic film or CCDs, the sensitivity of the eye was the limiting factor. As an example, the pupil diameter (aperture) of a dark-adapted eye is ~7 mm, and under good conditions can see down to the sixth magnitude unaided. A telescope with an objective diameter of 50 mm (2 inches) gathers $(50/7)^2 \approx 50$ times more light that the eye, or a little more than 4 magnitudes, so that an observer with this telescope could see as faint as tenth magnitude stars.

3.3 Older Detection Systems

In the first chapter, we briefly described the early magnitude scale and attempts in the nineteenth century to quantify and standardize it. The task was difficult because it relied on the human eye as a detector – its sensitivity varies from person to person and has a non-linear response to brightness and color. The parameters of the problem changed dramatically with the advent of photography, then the photoelectric tube, and finally the CCD chip. In this section, we briefly look at the older detectors, starting with photography. In the section following, we will examine the modern and ubiquitous CCD chip in some detail.

3.3.1 Photography

Photography became the detector of choice for measuring magnitude in the early twentieth century. Its advantages were numerous. Exposing a photographic plate to starlight while simultaneously guiding the telescope allowed starlight to build up, allowing fainter detections than possible with the human eye. The size and density of the star image on the photograph is a good proxy for the brightness of the star, so measurements of these properties provide a quantitative and repeatable way to measure magnitude. Uniform manufacturing processes meant that photographic plates could be standardized, avoiding the problem of visual differences between observers. And perhaps best of all, photographic plates could be preserved as a long-term record.

There were challenges. The precision of photometric photography was based on the assumption that the photometric plate behaved linearly. If exposed for twice as long, the image should be twice as large or twice as dark (on a negative). If two stars differed by a factor of two in brightness, their diameters or film density should differ by a factor of two. But early on, it became clear that these rules didn't hold as expected. Correction factors were required. There was also the problem of standardization. Even if manufactured at the same facility, one could expect some differences in sensitivity from plate to plate.

Photographic plates were also found to be more sensitive to blue light than the eye, so that blue stars looked brighter on photographs than to visual observers. This required the differentiation of two magnitudes – visual, m_v, and photographic, m_b (here the b refers to blue, while the v refers to visual). Although photography is no longer used by any professional observatory (and few amateurs), the major photometric systems in use (e.g. UBV) still employ a blue filter that was originally designed to mimic the response of film to light.

3.3.2 Photomultipliers

By the mid-twentieth century, the **photoelectric cell** was married to the telescope to fashion a photoelectric photometer, and later the photomultiplier photometer (Hearnshaw, 1996). A photoelectric cell emits electrons from a metal exposed to electromagnetic radiation, a process called the **photoelectric effect**. Those electrons generate a measurable current; the more light, the greater the current. A photomultiplier tube consists of a photoelectric cell in series with a number of electron multipliers. An incoming photon generates a few electrons which, upon being received by an electron multiplier, generate many more; these in turn strike another multiplier, and a cascade process rapidly and exponentially multiplies the total number of electrons, and therefore the measured current, from the incident photon. One of their great advantages over film is their linear response over a wide range of intensities. And as the technology advanced, they became far more sensitive than photographic plates and responsive to a variety of wavelengths. However, the device was bulky and required restricting diaphragms to exclude all light except that from the target star or planet. It was therefore useful for focusing on specific targets, like variable stars, but of limited value for larger surveys where the photographic plate still ruled.

3.3.3 Vidicons

For a limited time during the late middle twentieth century, some Earth-based photometric studies of extended objects, particularly of the Moon, were conducted by using an electronic vidicon tube in place of the photoelectric photometer (e.g. Shorthill and Saari, 1965). Vidicons are essentially cathode ray tubes. The planetary image was projected onto a screen at one end of the tube which was made of a photoconductive material, i.e. one whose conductivity changed with the intensity of light projected onto it. That surface was continuously scanned by an electron beam produced within the tube, and the detected changes in conductivity were proportional to the brightness on the screen at that spot. These early video cameras found their greatest photometric use on the first unmanned planetary spacecraft missions of the 1960s and 1970s, including NASA's Voyager and Viking missions (Thorpe, 1976; Birnbaum, 1982).

3.4 Charge-Coupled Devices

The modern **charge-coupled device** (CCD) camera merges the advantages of film, the photoelectric cell, and vidicon tube, and has very few disadvantages. A common analogy for a CCD is an array of closely spaced buckets that capture photons. After a given exposure period, the number of photons in each bucket is measured and the measurements are used to generate a digital image. Each light bucket is called a **picture element**, or **pixel** for short. The buckets are then emptied and readied for the next exposure where the process is repeated as desired.

There are three major advantages of the CCD over most of the previous recording methods. First, a CCD generates an electronic image that can be stored and transmitted anywhere, with no loss or degradation – it's just a set of numbers. Second, every object in the field of view is measured, allowing for immediate and direct comparison of visible stars and planets. Finally, they are extremely efficient. Film might require ten or more photons to begin to darken a light-sensitive grain on the emulsion, while the CCD can count individual photons.

In the 1970s–1990s, CCDs were rather small and limited to a few hundred thousand pixels. Since then, though, the technology has rapidly advanced and observatory quality CCDs have upwards of a billion pixels and unprecedented sensitivity over a wide range of wavelengths. There is an enormous literature just on CCD cameras for astronomy, and the technology is still rapidly changing. Below, we provide a very brief overview of these devices. An excellent and more detailed introduction can be found in Howell (2006).

3.4.1 Physical Characteristics

A charge-coupled device (CCD) is, in essence, a rectangular array of metal-oxide-silicate (MOS) capacitors – electrical devices that can store electrical charge. Each capacitor represents a pixel of a specific size, usually measured in micrometers × micrometers. As an example, the popular KAF-1001 CCD chip is a 1024 × 1024 array, and each pixel is 24 µm × 24 µm. Pixels need not be square, as here, but it is common.

The MOS capacitor is composed of three layers: an electrode attached to a metal gate (top), then a silicon dioxide insulator, then a doped-silicate semiconductor (bottom) attached to another electrode. The silicate is photosensitive, so incident photons of the proper wavelength knock electrons loose in the semiconductor layer via the photoelectric effect. Ordinarily, these would be lost, but the design of the MOS allows the loose electrons to be temporarily stored as a charge; each capacitor stores only the charge created by photons that impinged on its little area within the semiconductor. Over a wide range of lighting intensity, the CCD response is linear, so the amount of charge is directly proportional to the amount of incident light – up to the storage limit of the capacitor.

The **active image size** represents the sensitive area of the chip and is commonly measured in millimeters. The KAF-1001 chip documentation lists an active area of 24.6 mm × 24.6 mm. If one multiplies 1024 (the number of pixels per column) by 24 µm (the width of each pixel), one gets 24.6 mm – exactly the active width listed. The **fill factor** is the ratio of the active sensor area to the total area covered by the chip, so in this case, the KAF fill factor is 100%. Fill factors are reduced in some chip architectures (see Section 3.4.4).

The percentage of photons striking the chip that are recorded is given by the **quantum efficiency** (QE). This depends on the wavelength of light, so it is common to see a plot of QE versus wavelength in the reference literature for a sensor. For some detectors, as many as 9 out of 10 visible-wavelength incident photons can trigger a response, leading to a quantum efficiency of 90% or more. By comparison, film records only a few percent.

The earliest (and still common) layout for a CCD chip is **front-illuminated**; the gates are on the top and light must travel through and around them to trigger the photoelectric effect. This is, in large part, so that the silicate layer can be fairly thick and the chip mechanically robust. However, a significant fraction of light is absorbed before detection so that they may have efficiencies of *only* 50% or so. For more sensitivity, a **back-lit** or **back-illuminated** CCD reverses the layers so that the silicate is illuminated and the gates are on the bottom. This requires that the silicate layer be shaved thin, adding greatly to the expense and resulting in a more physically delicate chip. However, the efficiency of these chips can exceed 90% in some wavelengths, so the tradeoffs for high-end work are easily worth it.

Each pixel has a maximum electron storage capacity, known as the **full well capacity**, typically measured in kiloelectrons (ke^-). Generally, the larger the pixel, the more electrons it can store. The KAF-1001 chip has a full well capacity of 650 ke^-.

Like the analog bucket, once the pixel well is at maximum capacity, additional electrons will spill over into adjacent pixels, an effect referred to as **blooming**. This, of course, is undesirable because neither the full pixel nor the adjacent ones can now be trusted to give accurate counts. Some CCD architectures, referred to as **anti-blooming**, have been designed with specific overflow space to minimize the contamination of adjacent pixels, but one is still left with a saturated pixel that no longer accurately indicates the intensity of light. The extra overflow space also eats into the active space available for integration and reduces the overall quantum efficiency of a chip. These chips are therefore typically used only when the targets are relatively bright. If a chip is of the anti-blooming type, it will say so. If it is not, it is often unfortunately referred to with the double negative **non anti-blooming**.

3.4.2 Readout and Quantization

Once a CCD has been exposed to light, the charge stored in each pixel (capacitor) must be measured, a process referred to as the **readout phase**. The most straightforward way to do this would be to connect an output wire to each pixel, but this is impractical. Instead, engineers have devised a method to skillfully manipulate the voltage across each row of capacitors so that the stored charges are shifted down from row to row (in parallel) in a clock-timed sequence. After a row shift, the capacitors in the last row, alone, shift across sequentially (or serially) into an **output register**, a type of super pixel that can hold several times as many electrons as a single pixel. From there, the charge is *amplified* by on-chip electronics, and then sent to an **analog-to-digital** (A/D) converter, a device that measures the resulting voltage and converts it into an (integer) number (Figures 3.7 and 3.8).

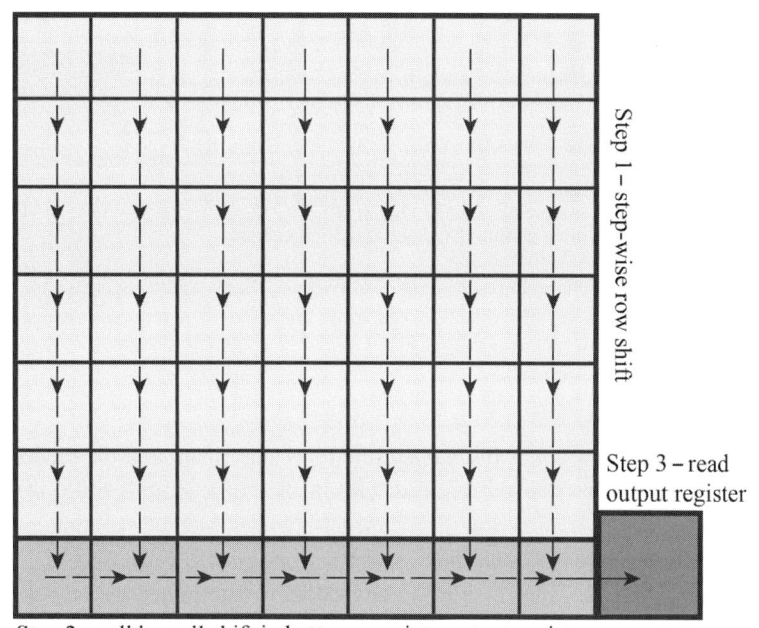

Figure 3.7. CCD readout sequence. After collection, the electron wells are shifted down one row at a time (step 1). After each shift, the cells in the bottom row are shifted right (step 2) where they are collected and read by the output register (step 3). If the cells are binned for observing, multiple cells are shifted into the output register prior to reading.
Credit: Michael K. Shepard.

Anti-blooming architecture

Output Register

Overflow channels collect and drain excess

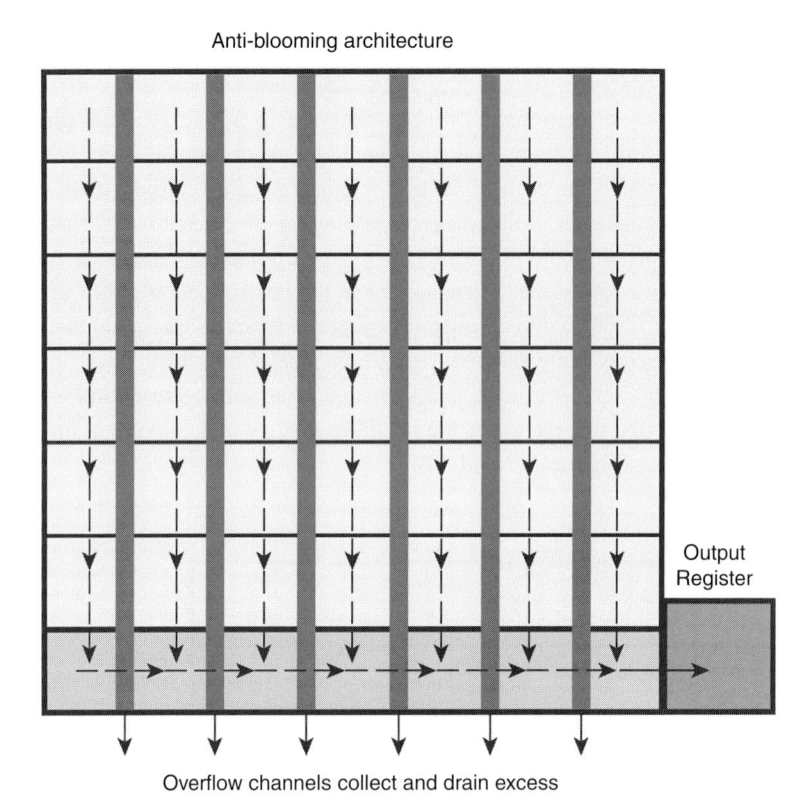

Figure 3.8. Anti-blooming architecture and readout. The overflow channels capture and drain excess electrons from the array. The trade-off is reduced area for collection and lower *SNR* characteristics.
Credit: Michael K. Shepard.

Once the entire row has discharged, the rows shift again and the process is repeated until all rows have been read. This method of readout is called **serial-parallel readout**, or sometimes the "bucket-brigade" method, continuing the analogy of pixels as buckets being passed from person to person in a fire brigade. The rate at which a chip can be readout is determined by a clock and its speed of operation is given in clock cycles per second (Hz). A typical CCD may have a 1 MHz clock, but this doesn't necessarily mean that it can read a million pixels in a second since each transfer and read may take several operations.

Even with fast clocks, the serial-parallel readout method takes time. To speed this up, some CCDs use a **multi-port readout** method. These shift rows up and down simultaneously (starting from the center) and have two sets of electronics measuring the pixel charges as each end row shifts. Or they may also split the end rows so that the one half shifts left and the other half shifts right, and use four sets of electronics (one at each corner) to measure the charges. The trade-off is that each set of measuring electronics differs from the others, and this may introduce systematic offsets in the final data that must be considered.

For a variety of reasons, the user may want to measure fewer pixels. It may be that they only want to read a subset of pixels; many CCDs offer this ability, a process referred to as **windowing**. This reduces the amount of data, important when transmitting images over the low bandwidth available to many spacecraft. Or perhaps the light source is dim or the seeing is poor and they want to increase the signal-to-noise ratio. In this case, many CCDs offer an option to **bin** pixels. Usually (but not always) binning takes place in powers of two, so one may bin 2×2 or 4×4 pixels into one. In this case, it is important to consider the maximum electron well depth of the output register.

After the output register has received the appropriate pixel charge (more than one pixel's worth in the case of binning), the charge is passed through an on-chip amplifier and converted to a voltage before being read by an analog-to-digital converter where it is quantized into in analog-to-digital units, or ADUs.[1] The number of electrons (or equivalent voltage) per ADU is referred to as the **gain** of the CCD.

At this point, a decision must be made – how many ADUs are needed? If one wishes to represent each pixel as a byte value (8 bits), then 256 (2^8) or fewer values can be used. Ideally, one would like to set the gain to maximize the dynamic range of the chip, so that 0 represents the minimum voltage expected (the noise floor) and 255 the full well depth. Although many cameras have used an 8-bit depth, it is now more common to see 10, 12, or 16 bits of ADU, representing ranges of 0–1,023 (2^{10}), 0–4,095 (2^{12}), and 0–65,535 (2^{16}), respectively. As an example, suppose the full well depth for a CCD is 15,000 e^-. If the gain of the electronics is 4 e^-/ADU, one needs 3,750 ADUs to represent a full well and 12 bits are needed.

Some cameras have multiple gain settings. Here, there are two or more amplifiers available to the chip. In the preceding example, one might choose between the standard 4 e^-/ADU gain state for normal operations and switch to a high gain setting of 1 e^-/ADU (paradoxically, high gain has the lowest e^-/ADU values) for dim objects.

---

[1] Spacecraft and other literature may use the term DN (digital number) in place of ADU if there is a 1:1 correspondence between the ADU and data used to generate the digital image.

3.4.3 Noise

Measuring photons with a CCD is subject to several kinds of noise. The most common measure of this is the **signal-to-noise ratio** (*SNR*), the ratio of the signal strength to the mean background level. To claim the confident detection of a signal against the background, most consider it necessary for *SNR* > 5.

For CCD applications, **shot noise** is independent of the optics or hardware. Instead, it is intrinsic to the nature of our measurements – we are counting a relatively small number of discrete photons. In one 10-second interval, we might count 1,000 electrons, while in the following 10 s, we might count 1,028, even though the source is exactly the same. For processes like these, the signal is subject to random variations that obey Poisson statistics and we report the **signal-to-noise ratio** (*SNR*) as

$$SNR = \frac{N}{\sqrt{N}} = \sqrt{N} \qquad 3.6$$

For a CCD, the maximum number of electrons per pixel is given by the full well capacity, so this sets the *best SNR* one could expect for a given measurement. For the KAF-1001, the full well capacity is 650 ke$^-$, so the best *SNR* for a measurement would be ~800.

If one simply runs through the readout process of a CCD without exposing the chip, the output will not be zero. Some constant amount of noise, called the **bias**, is intrinsic to each pixel, and each pixel will likely have a different bias value. To account for this, it is common to take several **bias frames** (images of zero exposure time) and average them to generate a master bias frame. This averaged frame captures the best bias estimate for each pixel in a CCD and is then subtracted from every subsequent image so that an exposure is starting from a true zero.

Dark current noise is created by thermal processes within the silicate and is given as the number of thermal electrons generated per pixel per second (e$^-$/pixel/s). Note that this differs from bias, which is for a zero-length exposure. At room temperatures, dark currents can be several thousand electrons per second. For ordinary digital cameras, this is acceptable given the large light levels in a typical daytime scene. For astronomical applications, however, this amount of noise often exceeds the signal of the intended target, so the detector must be cooled. Typical methods of cooling utilize some type of cooling medium (e.g. liquid N$_2$), thermoelectric cooling, or some combination. At temperatures of $-50°$ to $-100°$C, dark current can be made almost negligible, with values ranging from ~1 to fractions of an e$^-$/pixel/s.

To account for dark current, the CCD is exposed for the same amount of time as an image, but with the shutter closed so that the only accumulation is dark current. This is called a **dark frame** and is subtracted from the actual image to remove the effect. Note that the bias frame is intrinsic to the dark frame and need not be subtracted twice. A common practice on spacecraft CCDs is to permanently cover or **mask** one or more pixel columns to be used as a dark current reference.

Some of the signal in a pixel may be lost as the electrons are shifted over and over until finally read. Columns farthest from the output register will suffer more losses than those closest, so there may be a gradient in the output image referred to as **shading**. In modern CCDs, **transfer efficiencies** (listed with the CCD documentation) are quite good, so this is rarely a problem except for the largest arrays. The KAF-1001 transfer efficiency is given as >0.99997.

Readout noise is produced while the pixel charges are being amplified and read by the A/D electronics. Even if all of the electrons in a given pixel are transferred perfectly, their amplification to a larger voltage is an analog process subject to random fluctuations, again largely due to thermal processes. The speed at which the readout occurs comes into play because faster readouts generate slightly more heat in the electronics, meaning more readout noise. Finally, there is also error introduced by the very process of taking an analog voltage and digitizing it – each voltage is rounded to the nearest integer. Despite all these sources, readout noise has been reduced dramatically in most modern CCDs. The value of readout noise is provided by the chip manufacturer and is given in electrons (root-mean-square or rms) per pixel. The KAF-1001 CCD has a listed readout noise of 15 e^- rms/pixel when the readout rate is 1 MHz. This value is subtracted from every pixel.

A parameter listed for most CCDs, and closely related to the signal-to-noise ratio, is the **dynamic range**. This is the ratio of full well depth to readout noise:

$$dynamic\ range = \frac{full\ well\ depth}{readout\ noise} \qquad 3.7$$

It represents the maximum number of output ADU that one could expect to need (given no other noise sources), and generally constrains the maximum number of output bits needed. For the KAF-1001, full well depth is 650 ke$^-$/pixel and readout noise is 15 e^- rms/pixel, so the dynamic range is ~43,300 and could be fully digitized using 16 bits. This number is also reported in decibels (dB), given by

$$dynamic\ range\ (dB) = 20 \log \left(\frac{full\ well\ depth}{readout\ noise} \right) \qquad 3.8$$

The KAF-1001 therefore has a dynamic range of 93 dB.

Although it is not strictly a type of noise, every pixel in a CCD has a slightly different response to incident light; perhaps one pixel is 1% more efficient than another. **Flat fields** are images taken with the entire optical system to measure this pixel-to-pixel variation. The target is sometimes a white matte (Lambertian) screen illuminated evenly, or sometimes the twilight sky away from the rising or setting Sun. The assumption is that the screen or the sky at these times is uniformly illuminated, or flat, so variations in the output image are due to pixel-to-pixel variations. The flat field frame is normalized (the mean value is 1.0) and all images of targets are divided by the flat (after noise removal) to make the overall response uniform.

In all CCDs, there are inevitably misbehaved pixels. Some may not record at all – **dead pixels**. There are also **hot pixels** that show up as very bright, regardless of the exposure time. Usually, these are known and reported for each chip so they can be ignored. For spacecraft CCDs, exposure to the space environment will often degrade pixel performance, so these may change over time.

3.4.4 Architectures

The simplest architecture for a CCD is referred to as **full-frame**, and is essentially as described above; the chip is exposed for a set time followed immediately by pixel readout. The main problem with this design is that the pixels continue to accumulate photons as the columns are shifting for readout. This can result in smearing. The easiest fix for this is to incorporate a mechanical shutter, as with a traditional camera. The CCD is exposed while the shutter is open, and readout occurs only after the shutter is closed. This is generally the architecture of choice for astronomical cameras on ground-based telescopes.

In the **frame-transfer** architecture, half of the CCD is masked by an opaque layer. The uncovered half is exposed and then quickly transferred under the cover, where it is then read out at its (relative) leisure with no danger of additional exposure, and the uncovered half begins a new exposure. The disadvantage of this architecture is that only half of the CCD is used for actual imaging.

A variant on the frame-transfer theme is the **interline** architecture; in this design, each column of exposed pixels is adjacent to a covered row. After exposure, each column is shifted once into a covered column where it can be readout without additional exposure. This would also seem to have the same "waste" problem as the frame-transfer architecture (reducing the fill factor described earlier), but engineers have mitigated some of this by placing small lenses over the array to redirect (focus) light that would otherwise impinge on the mask into the exposed pixels instead. Both the full-frame and frame-transfer architecture are suitable for either front- or back-illuminated chips. However, the interline architecture can only be implemented (at least for now) with a front-illuminated chip.

In a spacecraft camera, if the design is to be of the framing camera type (as opposed to a pushbroom, discussed in the next section), the frame-transfer or interline architectures are usually preferred over the full-frame because mechanical shutters are prone to mechanical problems and difficult to fix when orbiting a distant planet. There is also the problem of smear during spacecraft close approach phases, especially in low light conditions. For these reasons, both the Long-Range Reconnaissance Imager (LORRI) on the New Horizons mission and the DAWN mission Framing Camera use a frame-transfer architecture (Cheng et al., 2008; Sierks et al., 2012).

3.4.5 Pushbrooms and Spectroscopy

The design we just discussed, common to all major telescopes and many spacecraft imaging systems, is referred to a **framing camera**; in essence, a CCD chip takes the place of film. A popular alternative for spacecraft imaging is the **pushbroom scanner**. In this system, the CCD is not an array, but a single line of pixels (e.g. 1×1024), oriented perpendicular, or **cross-track**, to the motion of the spacecraft. An image is built up as the spacecraft moves over the surface, one line at a time. A major advantage of this system is that it uses a simpler chip and readout electronics. However, the reconstruction of an image is more complex. The spacecraft motion may not be uniform (e.g. elliptical orbit) and the target planet is rotating. The tradeoff, then, is more complex processing and potential image distortion; one must know the attitude and motion of the satellite with high precision. Nevertheless, this is a popular mode for spacecraft and is used on the Mars Express High Resolution Stereo Camera, the LRO Narrow Angle Camera (NAC), and the Mars Global Survey Mars Orbiter Wide Angle Camera (WAC) (McEwen et al., 2007; Robinson et al., 2010; Figure 3.9).

A variant on the pushbroom is the **pushframe scanner**. In this configuration, the broom consists of more than a single array of pixels, e.g. $10 \times 1,024$, and the image is built by stitching together these **framelets** instead of single lines. Why do this? One reason is to subdivide a larger CCD array into separate frames, each of which is used by a specific optical or filtering system. For example, the LRO Wide Angle Camera (WAC) incorporates a single $1,024 \times 1,024$ CCD, but it is used by two different cameras (ultraviolet and a visible/near infrared), and framelets are covered by different color filters (Bowman-Cisneros and Eliason, 2011). This allows one CCD chip to be used to

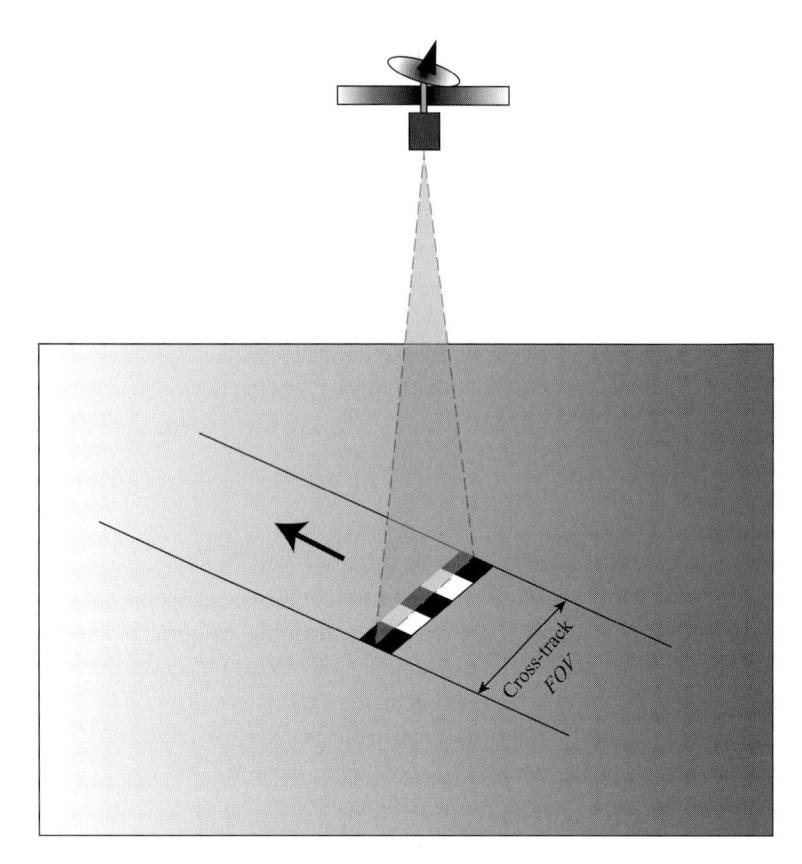

Figure 3.9. Pushbroom geometry. A CCD line array records the brightness of a narrow strip along its cross-track. An image is built up from a collection of these narrow strips as the array travels in the direction shown.
Credit: Michael K. Shepard.

simultaneously image the lunar surface in one monochrome, two ultraviolet, and five visible/near infrared bands with no moving parts.

Finally, although beyond the scope of this book, we should mention the role of CCDs in telescopic spectroscopy. One example is the SpeX spectrographic imager used at the NASA Infrared Telescope Facility (IRTF) (Rayner et al., 1998). In this optical system, light from an object of interest (e.g. an asteroid) passes through a slit and is either transmitted through a prism or reflected from a diffraction grating; the resulting dispersed spectrum is spread over and read by a CCD array as an image and post processing extracts the spectrum.

3.5 Measurement and Reduction

We've installed a CCD camera on a telescope or spacecraft camera and plan to make some observations. We will get some numbers from the CCD and want to turn them into photometric quantities – either absolute magnitudes or radiance units. In this section, we touch on the highlights of that processing pipeline and the physical variables that must be accounted for.

3.5.1 Matching the CCD and Telescope

The CCD camera and telescope work as a unit. In the first section, we described several optical properties of the telescope, including field of view and resolution. In any complete optical system, we must account for both the telescope and CCD to compute these quantities. Limiting magnitude is a bit more difficult to predict in advance because it depends upon many unknowns: the background sky brightness, seeing, exposure time, performance of the CCD, etc. For this reason, it is probably best found empirically for a few typical nights and exposure times. Then one can use the assumption of a linear response to approximate limiting magnitudes for other exposure times. For spacecraft systems, each camera undergoes an extensive calibration process before launch, and many include internal standard lamps or use other mechanisms (e.g. standard stars) to periodically check the calibration.

3.5.1.1 Field of View

The field of view is straightforward to compute. If we know the active size of the CCD, we can calculate the *FOV* using Equation 3.2. If the array is not square, there will be two different values. If the optical system is a pushbroom, the long dimension is the width and the *FOV* is reported as the **cross-track *FOV***.

One caveat to the field of view calculation is a phenomenon referred to as **vignetting** which refers to a progressive dimming as one moves away from the optical axis of the field of view. It is caused by two design issues in telescopes. In reflecting telescopes, the geometry of the tube length and mirror placement constrain the field of view. And virtually every folded telescope (e.g. Cassegrain) has a series of internal baffles to block stray light and makes a choice about how large a hole to put in the primary. The result is that, at a certain angular distance away from the optical axis, the amount of light available to the focal plane is reduced by partial blockage.

The second issue is the size of the secondary and tertiary mirrors. The larger these are, the more off-axis rays they capture, but the more incident light they block from the primary. The smaller they are, the fewer off-axis rays they collect from the primary reflection.

Every telescope will therefore have a specific maximum field of view that is fully illuminated. The calculation of that value is beyond the scope of this book, but may be found in the specifications of the telescope or computed using references in the bibliography. The important point to take away is that, when pairing a CCD camera with that telescope, it is best to match the camera *FOV* with that of the fully illuminated telescope *FOV* if possible. If it can't be done, don't despair; with the exception of a reduction in limiting magnitude near the field edges, most of the effects of vignetting can be reduced by the use of flat frames.

3.5.1.2 Resolution

Here is where the choice of CCD is critical for maximizing the telescope or camera optical performance. To make the discussion concrete, assume we are using the KAF-1001 CCD (discussed earlier) in our camera. It has 1024 × 1024 pixels, and each pixel is 24 µm × 24 µm. Let us attach this camera to a 0.1 m *f*/10 telescope; the focal length is therefore 1 m. At a distance of 1 m, a 24 µm pixel will subtend an angle of 24 µrad or 5 arcseconds. This number is referred to as the **plate scale** for this system. For spacecraft camera systems, it is more common to see it called the **instantaneous field of view** (***IFOV***; Figure 3.4). Under any circumstances, this is the best resolution we can expect from this system since the pixel is the smallest element in the image. However, the theoretical resolution of a 0.1 m telescope (Eq. 3.4 with λ = 550 nm) is 6.7 µrad or 1.4 arcseconds. This means that the camera is not capable of matching the resolution power of our telescope; the camera and telescope are poorly matched.

Consider, instead, a 1 m *f*/10 telescope with a 10 m focal length. In this system, the plate scale is 2.4 µrad or 0.5 arcseconds. The theoretical resolution of a 1 m telescope is 0.67 µrad or 0.14 arcseconds, well below this. At first, we might again think this is a poor match and that we should use a camera with smaller pixels. However, even the best observing sites in the world will rarely allow seeing better than 0.5 arcseconds; on that account, then, the telescope and camera would appear to be well matched. Almost.

Astronomers define the sampling parameter, r_{samp} as

$$r_{samp} = \frac{PSF_{FWHM}}{IFOV} \qquad 3.9$$

where PSF_{FWHM} is the full-width at half-maximum of the *PSF* in the same angular units as the *IFOV*. Sampling theory tells us that in order for one to distinguish and accurately measure a quantity like the point-spread function (*PSF*) of a star or asteroid, one must sample it at a resolution of at least half its size or smaller, i.e. $r_{samp} \geq 2$. This is referred to as **Nyquist sampling** (for **Harry Nyquist**, 1889–1976). In our example, the plate scale (or *IFOV*) is the same as the expected size of the *PSF*, 0.5 arcseconds, so $r_{samp} = 1$, a poor match. If, however, we use it at sites where the seeing is *no better* than 1 arcsecond, then we would meet the Nyquist sampling criterion and our counting statistics should be acceptable. But if we want to maximize the ability of the telescope when the seeing is exceptional, then we must either lengthen the focal length to 20 m (by whatever means), or use a camera with pixels that are 12 μm × 12 μm or smaller. A common pixel size in many CCDs is 9 μm × 9 μm, and these would give us an *IFOV* of 0.9 μrad or 0.2 arcseconds, for a sampling parameter $r_{samp} = 2.5$. Now the telescope and CCD are well matched.

There is a practical limit to how small the pixels should be compared with the *PSF*. If $r_{samp} > 3$, the *SNR* will suffer because each pixel will have fewer photons, but the noise per pixel (dark current, bias, and readout noise) will remain the same. The sweet spot for sampling is $r_{samp} = 2$–3 for the best statistics and *SNR*. However, given a choice, it is better to have r_{samp} too high than too low. One can always bin pixels in a CCD if necessary to adjust r_{samp}.

3.5.2 Observation and Processing

The processing and calibration of CCD data after acquisition is both art and science. A number of books have been written on the many details of the subject, and we recommend the excellent texts of Warner (2006) and Henden and Kaitchuck (1990). Here we touch only on the main points of reducing telescopic observations of a point source to a standard system, such as the Johnson & Cousins UBVRI (Chapter 2).

3.5.2.1 Calibration by Comparison

For telescopic observations, we calibrate the magnitude of our object by comparing it to other stars. If we are only interested in relative variations in our object, for example to generate a lightcurve, we only need to compare its output to one or more stars that do not vary. This is relatively straightforward and referred to as **differential photometry**. If, however, we wish to find absolute magnitudes – referred to as **absolute photometry** – we must compare our object with stars of known (or specified) magnitudes and perform a number of calibration steps.

Landolt stars (Chapter 2) were discussed as a common set of reference stars available to the astronomer. There are others available, but these have been examined thoroughly and are used by astronomers worldwide as the preferred standards for photometry. So in addition to measuring the object of interest, we must observe several reference stars, preferably all over the sky, using the same equipment on the same night. The images of our object and standard stars must be processed in the same way, using the general steps outlined in the following section.

3.5.2.2 Measuring the Object of Interest

For telescopic disk-integrated observations, there are at least two ways to do this. The most straightforward is called the **aperture method**. One encircles the object (or area) of interest on the image and counts all of the electrons within those pixels. Assuming a point source, the *PSF* will be (in a well-matched system) several pixels wide. A common rule of thumb is that a circle of radius $3 \times PSF_{FWHM}$ (in pixels) will capture all of the light for a point source (Howell, 2006). It is important to remember that the pixels will usually be in terms of DN (digital numbers) and must be converted to photon (or electron) counts by multiplying by the CCD gain. This is the baseline value from which to start.

The other, more complex measurement method assumes a model *PSF* which is fit to the observed data; this is **PSF profile fitting** method. The parameters of the best fit model *PSF* are then used to estimate the total signal count. Why do this? An early study of the Pluto–Charon system explains the need (Jones et al., 1988). High-resolution CCD images of Pluto–Charon were acquired with the Canada-France-Hawaii Telescope (CFHT) 3.6 m telescope on Mauna Kea, using a 640 × 1,024 RCA4 CCD with 15 μm × 15 μm pixels. The plate scale was 0.21 arcseconds / pixel, the seeing was on the order of 0.8–0.9 arcseconds, and the separation between Pluto and Charon was 0.93 arcseconds. In this case, the *PSF* of Pluto and Charon overlapped, so that counts from Pluto's *PSF* bled into Charon's *PSF* and vice versa. The aperture method could not separate the two signals. Instead, a separate *PSF* function was fit to each object so that contributions from the other could be accounted for. This provided a much more realistic estimate of their separate contributions to the signal.

3.5.2.3 Removal of Noise and Flat Fielding

A significant amount of effort goes into the generation of dark frames and flats (Section 3.4.3) and this is where it pays dividends. The dark frame (and intrinsic bias frame) is subtracted from the image, and the result is then divided (or multiplied, depending on how it is set up in the particular software) by the flat field. At this point, one has done the best job possible of removing known sources of CCD noise and leveling the response of the array.

There is another source of noise unrelated to the telescope or CCD – the background sky. Even with no stars visible, there is a buildup of photons on a CCD from intrinsic sky glow. This is exacerbated by light pollution, high clouds, or other atmospheric phenomena. To remove it, one must measure the CCD counts in pixels surrounding the target of interest (empty of stars!), get an average (or median) photon count per pixel, and subtract this value from each pixel containing the object of interest. Suppose the background sky averaged 20 e^-/pixel. Then if, for example, the object covered 25 pixels, and contained a total of 10,000 e^- (after dark and flat field corrections), one would subtract 25 × 20 = 500 e^-, giving a total of 9,500 electrons. The *SNR* would be $\sqrt{9,500}$ ~100 for this measurement.

3.5.2.4 Normalization

A common practice in many photometric software packages is to normalize the data to a standard exposure time, usually 1 s. If the CCD is linear (as expected), then the photon accumulation is directly proportional to the exposure time. The advantage of normalization is that it allows one to directly compare statistics of an object that may have been taken with different exposure times. If our example object was acquired with a 30 s exposure, we would normalize this to 9,500 e^-/30 s = 317 e^-/s.

3.5.2.5 Calculate Instrumental Magnitude

An **instrumental magnitude** is simply the magnitude computed if one uses the photon count for intensity in the magnitude equation.

$$m_{inst} = -2.5 \log(counts) \qquad 3.10$$

For our example, $m_{inst} = -6.25$ (using 317 for counts).

From this point on, the goal of the calibration is to find the "fudge" factors necessary to make the instrumental magnitude equal the true magnitude of the object. We must first remove the effects of the atmosphere.

3.5.2.6 Account for Extinction

Light from a star or target is attenuated as it passes through the atmosphere, and this is referred to as **first-order extinction**. Objects observed directly overhead travel through the shortest distance of atmosphere; those on the horizon, the most. Astronomers use the term **air mass** to describe the length of the travel path, in units that are *defined* to be 1.0 for objects at the zenith

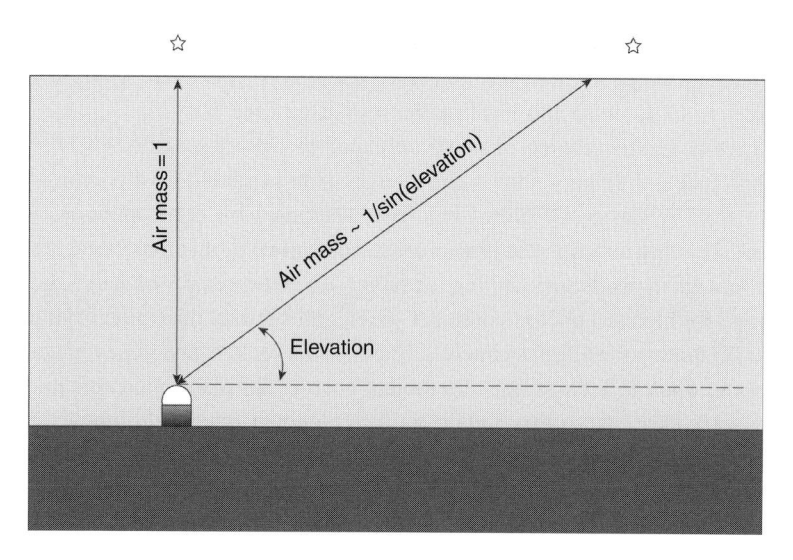

Figure 3.10. A telescope looks at a celestial object through the Earth's atmosphere. The closer the observation to the horizon, the greater the atmospheric depth, or air mass.
Credit: Michael K. Shepard.

(Figure 3.10). A brief look at the geometry of a travel path away from zenith demonstrates that a good approximation for the air mass is

$$air\ mass = \frac{1}{\sin(elevation)} \quad \quad 3.11$$

An object at the zenith has an elevation of 90° and an air mass of 1.0. At an elevation of 30°, the air mass is 2.0. This formula does not account for refraction of the light through the atmosphere or its radial density variations and slightly more complex formulae are typically used in practice (see Warner, 2006).

Extinction varies with the wavelength of the light. Shorter wavelengths, e.g. blue light, are attenuated more than longer wavelengths, e.g. red light. This wavelength variation is referred to as **second-order extinction**. As with first-order extinction, the greater the air mass, the larger the effect. It is usually negligible unless one is observing through a large air mass or there is a large difference in the color of the objects being examined.

3.5.2.7 Final Calibration

We will not go through the various steps of this process. It is involved and there are excellent references that explain it in great detail (e.g. Henden and Kaitchuck, 1990; Warner, 2006). However, it is worth examining the terms involved. The following equation gives the calibrated, or reduced standard magnitude, m_f, for a given filter type f, where f is a standard filter in the appropriate photometric system (e.g. Johnson–Cousins; Warner, 2006).

$$m_f = m_{inst,f} - k_f * airmass + T_f(CI) + Z_f \qquad 3.12$$

Here $m_{inst,f}$ is the (dark-corrected, flat-fielded, background sky removed) instrumental magnitude (from the counts), k_f is a constant that describes the rate at which light is attenuated with air mass, T_f is the **transform function** for the filter used, *CI* is the color index of the target (Chapter 2), and Z_f is called the nightly **zero point** for the filter.[2]

The air mass is found from the elevation of the target at the time of the observation. The constant, k_f, is found by observing a standard star over a range of air mass values (or standard stars at a variety of elevations) and using linear regression. The transform function corrects the optical system, including the filter, to better match the standard filter set. Filters will differ slightly in their absorption characteristics at different wavelengths, even if manufactured at the same facility, and every telescope and CCD differs from every other, so it is highly unlikely that any optical system will respond exactly as that used to define the standards. This function is analogous to the flats taken earlier – it corrects for those differences necessary to level all observations to a uniform standard. It depends the spectral color of the target, and that is most easily taken into consideration by incorporating its color index (e.g. $B - V$). Warner (2006) notes that, once the transform function is estimated for a telescope, camera, and filter combination, it changes little, if at all.

Finally, the zero point is the last correction factor. It takes into account all the other knowns and unknowns to bring the right side of the equation to the correct reduced magnitude. It changes from night to night (and if weather conditions are not stable, from hour to hour). By observing and reducing a number of known, standard stars, Equation 3.12 can be solved for k_f, T_f (if necessary), and Z_f. Once one is confident of these, they can be used to find the reduced magnitude of the previously unknown target.

3.5.3 Spacecraft Processing

For disk resolved observations, say for example a telescopic image of the Moon or spacecraft observation of an asteroid, one generally calibrates

---

[2] Unfortunately, zero point has two different meanings in the literature. Here, it refers to an instrumental term in photometric reduction; In Chapter 2, it refers to the standard calibration value (e.g. a particular star) for a given magnitude system.

individual pixels, although images may be smoothed or binned to remove noise, increase *SNR*, or reduce the processing time.

Spacecraft observations have several major differences from ground-based. There is no atmosphere (at least terrestrial) to worry about, thus no extinction. Seeing is not an issue, so matching the CCD plate scale and camera optics is more straightforward. Spacecraft color filters are generally custom made and not tied to any specific photometric system. Perhaps most differently, spacecraft observations are generally reported in units of radiance, not magnitudes. We will discuss this process in greater detail in Chapter 7.

4
The Physical Basis of Photometric Scattering Models

Scattering models are the mathematical relationships that tell us how much reflected light we can expect at any given illumination and viewing geometry. Early scattering models tended to be either simple theoretical or empirically-based relationships. More recent models build on these and attempt to account for the physical state of the photometric surface – its composition, roughness, etc. – with the goal of using the observed scattering behavior to extract and understand these properties. Before introducing these models, however, we first examine their physical basis.

4.1 The Nature of Light

4.1.1 Propagation and Notation

Light is a vector quantity, meaning it has both a magnitude and directional component. It is commonly conceived of as perpendicular electric and magnetic fields that travel in a direction perpendicular to both fields so that it is a transverse wave. The two waves oscillate in phase with each other and can be thought of as mutually reinforcing one another. In a vacuum, the propagation of a light wave can be written in several ways including

$$\mathbb{E}(z,t) = A \sin(kz - \omega t) = A \sin \varphi \qquad 4.1$$

or

$$\mathbb{E}(z,t) = A\, e^{i(kz - \omega t)} = A\, e^{i\varphi}$$

where $\mathbb{E}(z,t)$ is the amplitude of a propagating electric field wave of **maximum amplitude**, A, k is the **wavenumber** ($2\pi/\lambda$), z is the distance along the direction of travel, ω is the **angular frequency** (2π/wave period), t is time, and we commonly group the parameters in parentheses into a single parameter called the **phase** of the wave, φ.

The second wave equation uses complex number notation, where $i = \sqrt{-1}$, instead of the sine. This is often done because it simplifies the mathematics of accounting for multiple waves. Similar equations can be written for the magnetic field.

4.1.2 Polarization

As noted, light can be thought of as a transverse wave – one that is composed of a pair of perpendicular electric and magnetic fields propagating in a direction perpendicular to them both. If the electric and magnetic field vectors define an x–y plane, then the z-axis is along the direction of travel. We use the electric field vector to define another wave property called its **polarization**.

By convention, we say that the **sense of polarization** of a light wave is the same direction as that of its dominant electric field component. **Natural light** – that produced by emission from stars, electric lamps, and flames – is **unpolarized**. This means that the individual sources (atoms or molecules) emit their light with randomly oriented electric fields so that there is no dominant direction of the aggregate electric field vectors. If, however, one sense of polarization is preferred more than others, we say the light is **partially polarized**. If the sense of polarization is the same for all of the light, we say it is **completely** or **fully polarized**.

Light that is completely polarized within a single plane is referred to as **plane** or **linearly polarized**. For ease of visualization, consider light traveling along a direction parallel to the ground or other suitable plane of reference toward the viewer. By convention, we say that if the light (electric field) is oriented parallel to the ground plane it is **horizontally polarized** (0° linear polarization), while light oriented vertically with respect to this plane is **vertically polarized** (90° linearly polarized; Figure 4.1).

Consider two light waves of the same amplitude, traveling in the same direction, but polarized perpendicularly with respect to each other. If the waves are in phase with one another, the field vectors add so that the plane of polarization is now between the two individual planes. If the waves are 90° out of phase with each other, the polarization vector appears to rotate in a circle – **circular polarization** – as first one and then the other field dominate. If the two waves are any other amount out of phase, or if the two waves have different amplitudes and are out of phase by any amount, the polarization vector appears to rotate in an ellipse – **elliptical polarization**.

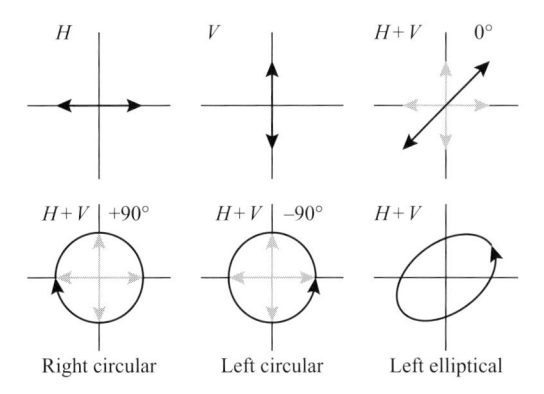

Figure 4.1. Illustration of the different states of polarization.
Credit: Michael K. Shepard.

4.1.3 Jones Vectors

If a light source is completely polarized, it can be represented succinctly using **Jones vectors**, named for the American physicist **R. Clark Jones** (1916–2004). The use of these vectors with matrix elements to be discussed later is called the *Jones calculus*. In this notation, a polarized wave is mathematically described as the superposition of horizontally and vertically polarized plane waves of magnitude A_h and A_v and relative phases $e^{i\varphi_h}$ and $e^{i\varphi_v}$. The generic Jones vector for a polarized light wave can then be written

$$\begin{pmatrix} A_h e^{i\varphi_h} \\ A_v e^{i\varphi_v} \end{pmatrix} \qquad 4.2$$

For light of amplitude $A = 1$, we write the Jones vectors for the cases described in the previous section as

$$\underbrace{\begin{bmatrix} 1 \\ 0 \end{bmatrix}}_{\text{horizontal}} \quad \underbrace{\begin{bmatrix} 0 \\ 1 \end{bmatrix}}_{\text{vertical}} \quad \underbrace{\begin{bmatrix} 1 \\ 1 \end{bmatrix}}_{\text{linear}+45°} \quad \underbrace{\begin{bmatrix} 1 \\ -1 \end{bmatrix}}_{\text{linear}-45°} \quad \underbrace{\begin{bmatrix} 1 \\ i \end{bmatrix}}_{\text{left circular}} \quad \underbrace{\begin{bmatrix} 1 \\ -i \end{bmatrix}}_{\text{right circular}}$$

The i in the last two matrices indicate that the phases of one of the two waves either lags or leads the other wave by 90° ($\pi/2$) to give circular polarization.

To see the usefulness of Jones vectors, if we add a horizontal and vertical wave of equal amplitude, A, we get

$$A \begin{bmatrix} 1 \\ 0 \end{bmatrix} + A \begin{bmatrix} 0 \\ 1 \end{bmatrix} = A \begin{bmatrix} 1+0 \\ 0+1 \end{bmatrix} = A \begin{bmatrix} 1 \\ 1 \end{bmatrix} \qquad 4.3$$

a linearly polarized wave of amplitude A oriented $45°$ from horizontal. If we add left and right circularly polarized waves of equal amplitude, we get

$$A\begin{bmatrix} 1 \\ i \end{bmatrix} + A \begin{bmatrix} 1 \\ -i \end{bmatrix} = 2A \begin{bmatrix} 1+1 \\ i-i \end{bmatrix} = 2A \begin{bmatrix} 1 \\ 0 \end{bmatrix}$$

which is a horizontally polarized wave of amplitude $2A$.

If we insert an optical element in the path of a light source, we can use Jones vectors to find the output light properties using a **Jones matrix** for the optical element. For example, a linear polarizer designed to transmit horizontally polarized light has a Jones matrix of

$$\begin{bmatrix} 1 & 0 \\ 0 & 0 \end{bmatrix}$$

We can find the effect of this filter on horizontally incident polarized light of amplitude $A = 1$ by pre-multiplying

$$\begin{bmatrix} 1 & 0 \\ 0 & 0 \end{bmatrix} \begin{bmatrix} 1 \\ 0 \end{bmatrix} = \begin{bmatrix} 1 \\ 0 \end{bmatrix}$$

and on vertically incident light as

$$\begin{bmatrix} 1 & 0 \\ 0 & 0 \end{bmatrix} \begin{bmatrix} 0 \\ 1 \end{bmatrix} = \begin{bmatrix} 0 \\ 0 \end{bmatrix}$$

In the first case, the filter allows all of the horizontally polarized incident light through. In the second case, none of the vertically polarized light passes through the horizontal filter.

A *quarter-wave plate* is a birefringent material (like quartz) cut in such a way so as to introduce a $\pi/2$ phase shift between different senses of polarization within incident light. We can write the Jones calculus for the introduction of a quarter-wave plate into the path of light of unit amplitude that is linearly polarized at $45°$ as

$$\begin{bmatrix} 1 & 0 \\ 0 & i \end{bmatrix} \begin{bmatrix} 1 \\ 1 \end{bmatrix} = \begin{bmatrix} 1 \\ i \end{bmatrix}$$

to show that a quarter-wave plate will change linearly polarized light into circularly polarized.

4.1.4 Stokes Parameters

The Jones vectors and calculus described in the previous section are only for fully polarized light. A more general system of parameters and mathematics for light in any polarization state uses the **Stokes parameters**, named for the Irish mathematical physicist Sir **George Stokes** (1819–1903).

There are four Stokes parameters, often designated I, Q, U, and V, although sometimes one will see these written as S_0, S_1, S_2, and S_3. These four parameters make up a **Stokes vector**, $\overline{S}$, and together completely describe the polarization state of a light wave.

$$\overline{S} = \begin{bmatrix} I \\ Q \\ U \\ V \end{bmatrix} = \begin{bmatrix} S_0 \\ S_1 \\ S_2 \\ S_3 \end{bmatrix} \qquad 4.4$$

A discussion of these parameters in any detail is beyond the scope of this book but may be found in references such as Fowles (1975). Here we limit our discussion of these parameters to the following. The parameter I is a measure of the total intensity of the light beam. The amount of energy in a beam of light is proportional to the square of the electric field strength, so one will commonly see this parameter written as

$$I = \langle \mathbb{E}_h^2 \rangle + \langle \mathbb{E}_v^2 \rangle \qquad 4.5$$

where the brackets indicate the expectation (or average) value of the value.

The parameters Q, U, and V are measures that describe the degree and orientation of the polarization sense. Some examples may help give a sense of how these work within a Stokes vector to describe a particular light wave.

$$\begin{bmatrix} 1 \\ 0 \\ 0 \\ 0 \end{bmatrix} \quad \begin{bmatrix} 1 \\ 1 \\ 0 \\ 0 \end{bmatrix} \quad \begin{bmatrix} 1 \\ -1 \\ 0 \\ 0 \end{bmatrix} \quad \begin{bmatrix} 1 \\ 0 \\ 0 \\ 1 \end{bmatrix} \quad \begin{bmatrix} 1 \\ 0 \\ 0 \\ -1 \end{bmatrix} \quad \begin{bmatrix} 1 \\ 0 \\ 1 \\ 0 \end{bmatrix}$$
unpolarized horizontal vertical right circular left circular +45° linear

Analogous to the Jones calculus, the Stokes vectors can be used with a 4×4 matrix that represents an optical element to determine its effect on the incident light. This matrix is called a **Mueller matrix**, after the Swiss-American physicist **Hans Mueller** (1900–1965), and the mathematical use of these matrices with Stokes vectors is called the *Mueller calculus*.

4.2 Refraction, Fresnel Reflection, and Absorption

In this section, we introduce the ways in which light interacts with matter, generically referred to as **scattering**.

4.2.1 Refraction

If a beam or ray of light is incident on a plane surface at an angle i from the normal (the incidence angle), a fraction of that light is reflected away at the same angle so that the emission angle $e = i$. No light will be observed at any other emission angle. The fraction of light not reflected is transmitted into the plane surface, or **refracted** at an angle i'. The amount of refraction depends upon a property of the medium known as the **index of refraction**, n, which can be defined as the ratio of the speed of light in vacuum, c, to that in the medium, v:

$$n_\lambda = \frac{c}{v} \qquad 4.6$$

We have included the subscript λ here to make explicit that the index of refraction is a function of the wavelength of light, but we leave it as an implicit variable in the following equations. When light enters a medium other than vacuum, it slows down. By inspection, it is clear that $n = 1$ for a vacuum ($v = c$), and $n > 1$ for all other *natural* substances. It should be noted that scientists have engineered materials, called **metamaterials**, for which $n < 1$; these materials do not transmit light faster than c, but cause light to refract in unnatural ways.

For a plane surface of refractive index n, the refraction angle i' is given by **Snell's Law**:

$$\sin i' = \frac{n_1}{n_2} \sin i \qquad 4.7$$

where n_1 is the refractive index of the media with the incident wave (often assumed to be a vacuum with $n_1 = 1$) and n_2 is the index of the media of refraction. The law is named for the Dutch astronomer **Willibrord Snell** (1580–1626), although there is good evidence it was discovered centuries earlier by the Persian physicist **Ibn Sahl** (Rashed, 1990). Because n is a function of wavelength, white light is spread into separated colors after passing through many media; this is referred to as **dispersion**.

4.2.2 Birefringence

Some materials are optically anisotropic, meaning that their refractive index varies with the state of polarization and the direction of light propagation relative to optical axes intrinsic to the material. These are called **birefringent** materials, and as pointed out in Chapter 1, were instrumental in the recognition that light has the property of polarization. The property was first recognized in crystals of calcite by Bartholin in the seventeenth century, and is the basis for the identification of many minerals in the modern optical microscope. It can also be induced by stress and, in some cases, by applied electric or magnetic fields.

Some optical devices, such as the quarter-wave plate described earlier, take advantage of the differing indices of refraction for different polarization states to induce phase shifts in the different states of polarization, changing, for example, linearly polarized light into circularly polarized.

4.2.3 Fresnel Reflection

The simplest case of reflection familiar to all is specular, Fresnel, or mirror-like scattering (Figure 4.2). Here one is dealing with a surface many times larger than the light wavelength, λ, in horizontal extent, and plane and smooth at the scale of the wavelength. In practical terms, this means that the vertical displacement from a plane everywhere is on the order of $\lambda/4$ or smaller.

The fraction of incident light reflected and refracted depends upon the incidence angle and the optical properties of the two media at the reflection boundary. For non-magnetic materials, which is a reasonable approximation for most planetary media (save perhaps metal-rich asteroid regolith), the reflected fraction, $r_{Fresnel}$, is given by the **Fresnel equations**, named for the French physicist Augustin-Jean Fresnel (Chapter 1). For unpolarized incidence light:

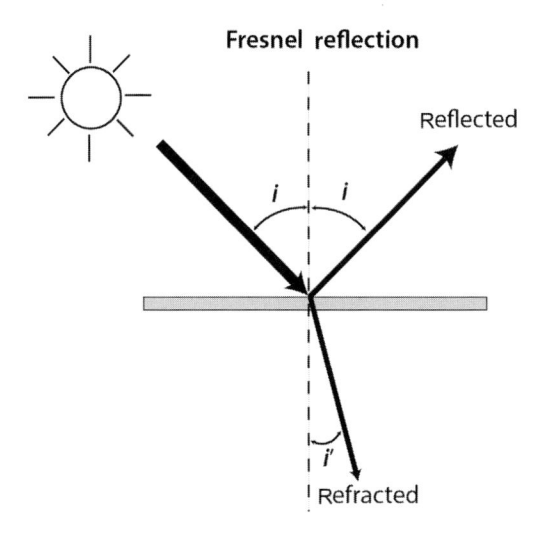

Figure 4.2. Fresnel reflection and refraction.
Credit: Michael K. Shepard.

$$r_{Fresnel} = \frac{1}{2} \left[\frac{n_1 \cos i - n_2 \sqrt{1 - \left(\frac{n_1}{n_2} \sin i\right)^2}}{n_1 \cos i + n_2 \sqrt{1 - \left(\frac{n_1}{n_2} \sin i\right)^2}} \right]^2$$

$$+ \frac{1}{2} \left[\frac{-n_2 \cos i + n_1 \sqrt{1 - \left(\frac{n_1}{n_2} \sin i\right)^2}}{n_2 \cos i + n_1 \sqrt{1 - \left(\frac{n_1}{n_2} \sin i\right)^2}} \right]^2 \qquad 4.8$$

Here, n_1 is the index of refraction for the initial media, and n_2 is the index of refraction for the new boundary media. For most planetary applications, n_1 is assumed to be 1.0, exactly correct for a vacuum and approximately correct for a thin atmosphere, and $n_2 > 1.0$. The refracted, or transmitted fraction, T, is simply $T = 1 - r_{Fresnel}$.

This rather complex looking equation is due to the fact that unpolarized incident light is partially polarized upon reflection from a smooth interface. The first term gives the reflection coefficient for the horizontal component and the second for the vertical component. Up until now, we have used the terms horizontal and vertical polarization – these are the terms used by radar astronomers. In optical astronomy, however, there is a change in terminology. The horizontal component (first term) is called the **perpendicular** component, so called because it is perpendicular to the scattering plane made of the incident and reflected ray (and thus horizontal to the surface), and the vertical component (second term) is called the **parallel** component of polarization. One will often see terms with the subscripts $\perp$ and $\parallel$ to indicate perpendicular and parallel components.

In the event that the incidence angle is zero (perpendicular to the surface), the equations simplify to

$$r_{Fresnel} = \left(\frac{n_2 - n_1}{n_2 + n_1}\right)^2 \qquad 4.9$$

and there is no polarization in the reflected waves. As an example, at optical wavelengths common glass has $n_2 = 1.5$, and air has $n_1 \sim 1.0$. For light perpendicularly incident on a thick sheet of glass, approximately 4% is reflected and 96% is transmitted.

For all other cases, there is always some amount of polarization in the reflected light and the reflection coefficient increases with incidence angle. Consider again light incident on to glass. A plot of the two components of reflection is shown in Figure 4.3 (excluding the 0.5 coefficient in front

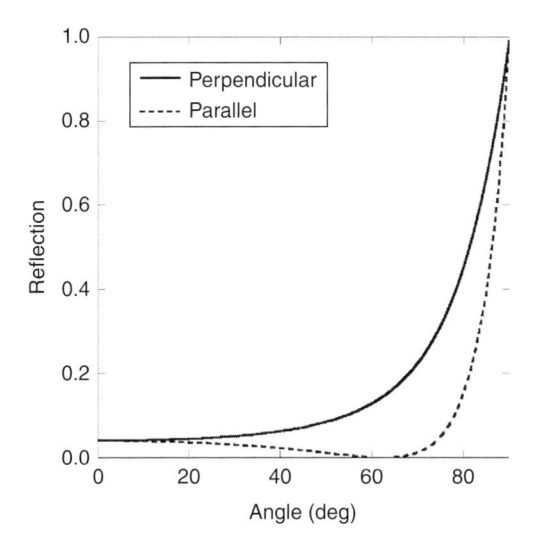

4.2 Refraction, Fresnel Reflection, and Absorption

Figure 4.3. Fresnel reflection components (perpendicular and parallel, Eq. 4.8) with angle for a surface of $n = 1.5$. At $i = 63°$, the parallel polarization goes to 0. This is the Brewster angle.
Credit: Michael K. Shepard.

of each term). Note that while the perpendicular (horizontal) polarized component is always greater than zero, the parallel (vertical) polarized component equals zero at one particular angle, called the Brewster angle, i_b:

$$i_b = \tan^{-1} \frac{n_2}{n_1} \qquad 4.10$$

At that angle, there is no reflection of vertically polarized light – only transmission into the glass.

There are only a few cases in planetary photometry when Fresnel scattering is applicable. Recent images of Saturn's moon Titan show specular glints off of smooth hydrocarbon seas (Figure 4.4), and it might yet be seen on smooth icy surfaces on some outer moons. But the smooth surface condition is too restrictive for it to be of general applicability to most planetary objects. It is, however, of value for modeling some types of scattering behavior, such as the negative polarization observed in asteroids (Geake et al., 1984; Chapter 6).

4.2.4 Absorption

When light enters a medium other than vacuum, it is absorbed or attenuated, at least to some degree. The amount of absorption is often specified by the complex index of refraction, which we write as

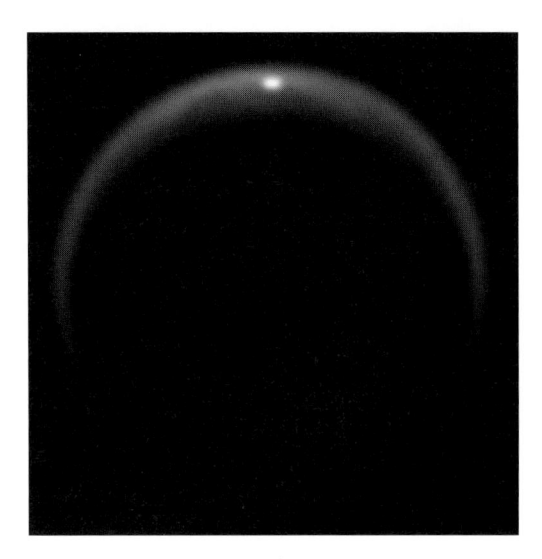

Figure 4.4. A specular glint of sunlight from the smooth ocean surface on Titan. Credit: NASA/JPL/University of Arizona/DLR. www.nasa.gov/mission_pages/cassini/multimedia/cassini20091217.html

$$\boldsymbol{n} = n + i\zeta \qquad 4.11$$

Here n is the real part of the index of refraction (shown earlier), and ζ is the imaginary part of the index and referred to as **extinction** or **absorption coefficient**.[1] To see how this leads to absorption, we write the electric field vector of a plane wave passing through a medium using its complex form:

$$\mathbb{E}(z,t) = Re\left[Ae^{i(kz-\omega t)}\right] \qquad 4.12$$

where the angular wave number ($k = 2\pi\boldsymbol{n}/\lambda$) incorporates the complex index of refraction. Substituting for this in the angular wave number and rearranging gives

$$\mathbb{E}(z,t) = A\left(e^{-\frac{2\pi z \zeta}{\lambda}}\right) Re\left[e^{i(kz-\omega t)}\right] \qquad 4.13$$

The term in parentheses shows that the amplitude of the wave is exponentially attenuated with distance traveled. The intensity of the wave (power) is

---

[1] Many authors use κ, kappa, for this parameter. We have chosen to use ζ, zeta, to avoid confusion with another extinction parameter in Section 4.5.2.

proportional to the square of the electric field, so the intensity falls off as $\exp(-4\pi z \zeta/\lambda)$ as the wave travels through the medium.

4.3 Scattering by Particles

4.3.1 Size Parameter

In many models, we assume the scattering medium is composed of individual particles that each contribute to the overall scattering behavior of the entire medium. One of the most important properties for predicting scattering behavior of each particle is its size. This may range from the molecules making up gases in an atmosphere, to droplets in clouds, to larger particles making up a planetary regolith. To conveniently account for the size of the particle, it is often normalized with respect to the wavelength of incident light, giving us the **size parameter**, X:

$$X = \frac{2\pi s}{\lambda} \qquad 4.14$$

Here, s is the average radius (size) of the particle. In general, scattering is discussed as belonging to one of three regimes: $X \ll 1$ or Rayleigh scattering, $X \sim 1$ or Mie scattering, and $X \gg 1$ or geometric optics.

The advantage of normalization in this way is that scattering behaviors can be easily scaled and compared. For example, molecules are on the order of 10^{-9} m in size and visual wavelengths 10^{-6} m, giving a size parameter $X \sim 10^{-3}$. Similarly, sand grains are on the order of 10^{-4} m and microwaves are 10^{-1} m, giving a similar size parameter. Therefore, we might expect microwaves to scatter from dispersed sand grains in a manner similar to that of light from atmospheric molecules.

4.3.2 Rayleigh Scattering

If the particles are much smaller than the incident wavelength so that $X \ll 1$, they scatter according to a mechanism known as **Rayleigh scattering** after the British physicist **John W. Strutt** (1842–1919), better known as **Lord Rayleigh**. Particles of this scale (typically atoms or molecules of a gas) resonate with the electric field of the incident wave, and re-radiate the incident light as a dipole antenna with a characteristic pattern (Figure 4.5). The intensity of scattering is an inverse function of wavelength (λ^{-4}), such that smaller size parameters scatter more. This is why the sky appears blue – blue (i.e. shorter) wavelengths scatter far more efficiently than longer ones, like red.

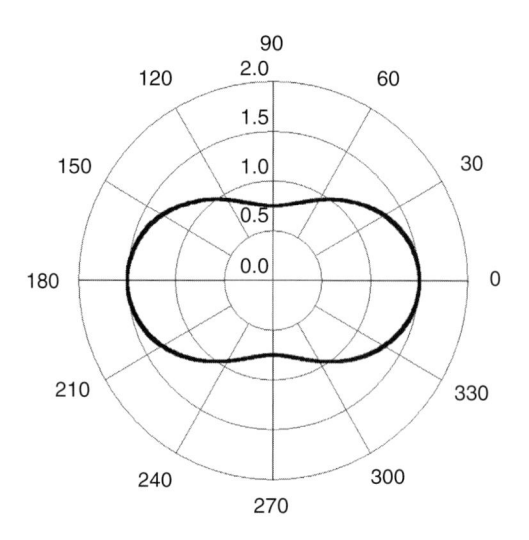

Figure 4.5. The particle phase function of Rayleigh scattering. The angle shown is the phase angle; 180° is forward scattering, 0° is backscattering.
Credit: Michael K. Shepard.

4.3.3 Mie Scattering

A stickler for detail will tell you that any scattering from any spherical particle is a type of **Mie scattering**, named for the German physicist **Gustav Mie** (1869–1957). Mie solved Maxwell's equations for the scattering of an electromagnetic plane wave from a homogeneous sphere; this solution is also sometimes referred to as the **Lorenz-Mie solution** because the Danish physicist **Ludvig V. Lorenz** (1829–1891) independently solved the same problem. The Mie equations allow one to predict how much light is scattered (the scattering cross-section), its intensity, polarization, and directional dependence for a homogeneous sphere of any size parameter, X.

For small size parameters, the Mie equations reduce to the equivalent Rayleigh scattering solution. For large size parameters, they lead to behavior well described by geometric optics (Section 4.3.4). For size parameters near $X = 1$, the scattering behavior can vary considerably and exact solutions often require numerical methods. It is for size parameters near $X = 1$ that most users imply when referring to *Mie scattering*.

While the atoms and molecules of an atmosphere are small enough to use the Rayleigh scattering approximation to the Mie equations, the non-gaseous constituents within an atmosphere – dust, droplets, and other particulates – are often close to the wavelength in size ($X \sim 1$). In other fields and in industry, the Mie solutions for $X \sim 1$ are important for understanding the scattering behavior of light within colloidal suspensions (insoluble microscopic particles dispersed throughout a liquid or gel) like milk and paint. The solutions to these scattering problems are beyond the scope of this book and the interested reader is referred to classical references such as van de Hulst (1981) and Bohren and Huffman (1983).

4.3.4 Geometric Optics

For the surfaces of planets, it is often assumed that the particles making up the regolith are large compared to the wavelength of visible light, so that $X \gg 1$. This is called the **geometric optics** approximation and allows us to treat incident light as if it is composed of individual rays that travel in straight lines until crossing an interface with a material of different optical properties (e.g. refractive index). In this case, the scattering cross-section for a particle is closely related to its actual, geometric cross-section, and the concepts of Fresnel reflection, refraction, absorption, and diffraction can be used to predict the intensity and direction of scattering. Most of the models we will discuss make this approximation in their theoretical underpinnings.

4.4 Particle Phase Function

4.4.1 Definition

An important property of a scattering particle is its **particle phase function**, P. This is a function that describes the angular distribution of scattered light. It is commonly assumed that the angular distribution is symmetric about the line of travel. While this may not be true in general for non-spherical particles, it can be safely assumed if the media consists of similar particles in a random orientation – as in the upper few microns of a planetary regolith. In this case, the random orientation smooths out the scattering asymmetry; the probability that light will scatter at a **scattering angle** θ is then given by $P(\theta)$. Note that the phase angle, α, is the supplement of θ ($\theta + \alpha = 180°$; Figure 4.6).

Since P is essentially a probability function, it must normalize over all scattering angles. This can be expressed as

$$\frac{1}{4\pi}\int_0^{4\pi} P(\theta)d\Omega = 1 \qquad 4.15$$

We can rewrite the solid angle in terms of scattering angle θ giving

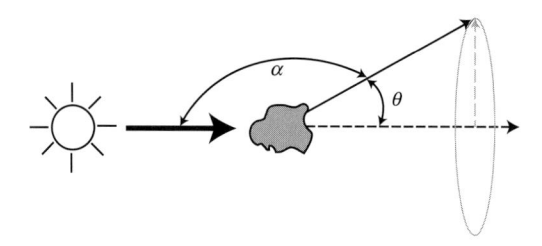

Figure 4.6. Particle phase function geometry. The phase angle is α and the scattering angle is θ.
Credit: Michael K. Shepard.

$$\frac{1}{4\pi}\int_0^\pi P(\theta) 2\pi \sin\theta d\theta = \frac{1}{2}\int_0^\pi P(\theta)\sin\theta d\theta = 1 \qquad 4.16$$

In most cases, we will express the phase function in terms of phase angle, α, instead of scattering angle. The normalization condition is identical – simply replace θ with α in the equation.

4.4.2 The Asymmetry Factor

Scientists are always looking for ways to simply describe something that is complex. With a particle phase function, we can provide a single number, called the **asymmetry factor** or **asymmetry parameter**, ξ (Greek xi)[2] which gives a general sense of the dominant direction of scattering: does is scatter more in the forward direction? Backward direction? Or symmetrically about both directions? The asymmetry factor is an average cosine weighting of the phase function, defined by

$$\xi = \overline{\cos\theta} = \frac{1}{2}\int_0^\pi P(\theta)\cos\theta \sin\theta d\theta \qquad 4.17$$

or because $\alpha = \pi - \theta$

$$\xi = -\overline{\cos\alpha} = -\frac{1}{2}\int_0^\pi P(\alpha)\cos\alpha \sin\alpha d\alpha \qquad 4.18$$

---

[2] Hapke (2012) uses the Greek letter ξ (xi) for the asymmetry factor and we follow that convention.

For particles that dominantly scatter in the forward direction, $\zeta > 0$; for those that are dominantly backscattering, $\zeta < 0$; for those that scatter symmetrically, and this includes those that scatter uniformly, $\zeta = 0$.

4.4.3 Isotropic Phase Function

The simplest case of scattering is one in which all directions are equally probable.

$$P(\alpha) = 1 \qquad 4.19$$

For this case, the asymmetry factor $\zeta = 0$.

4.4.4 Rayleigh Phase Function

Rayleigh scattering, recall, assumes particles much smaller than the incidence wavelength of light ($X \ll 1$). In this case, the molecules act as electric dipoles and the scattering phase function that results is (Figure 4.5)

$$P(\alpha) = \frac{3}{4}\left(1 + \cos^2\alpha\right) \qquad 4.20$$

Because the phase function is symmetric about $\alpha = \pi/2$, the asymmetry factor is again $\zeta = 0$.

4.4.5 Legendre Polynomial Phase Functions

A once commonly used phase function is the one- or two-term Legendre polynomial. A zeroth-order Legendre polynomial is the same as the isotropic phase function. The first- and second-order Legendre polynomial phase functions are given by

$$P(\alpha) = 1 + b_1 \cos\alpha \qquad 4.21$$
$$P(\alpha) = 1 + b_1 \cos\alpha + b_2 \left(3\cos\alpha^2 - 1\right)$$

where b_1 and b_2 are constants, and they are restricted in range to ensure that $P(\alpha)$ is normalized (Hapke, 2012).

4.4.6 Henyey–Greenstein Phase Function

In 1941, the astronomers **Louis G. Henyey** (1910–1970) and **Jesse L. Greenstein** (1909–2002) were attempting to verify the existence of a diffuse interstellar radiation, which required them to account for all light sources in their measurements of the brightness of the sky. They utilized a model of radiative transfer and introduced a new phase function to account for the directional nature of light scattered from interstellar matter:

$$P(\theta) = \frac{1 - \xi^2}{\left(1 - 2\xi \cos\theta + \xi^2\right)^{3/2}} \qquad 4.22$$

or

$$P(\alpha) = \frac{1 - \xi^2}{\left(1 + 2\xi \cos\alpha + \xi^2\right)^{3/2}}$$

This function is widely referred to as the **Henyey–Greenstein** (H-G) phase function and the parameter $-1 \leq \xi \leq 1$ determines whether the function is generally forward ($\xi > 0$) or back ($\xi < 0$) scattering (Hapke, 2012). Figure 4.7 shows an example for forward scattering ($\xi = 0.3$). Note that even though these functions are dominated by scattering in either a forward or backward direction, there remains a small component of the opposite direction. Hapke (2012) points out the many advantages of this function, including that it is normalized and the asymmetry parameter (Eq. 4.18) is the variable in the function. For $\xi = 0$, the function is isotropic, while for $\xi = \pm 1$, the function is a delta function in the appropriate scattering direction.

Two H-G functions are often used in tandem to model scattering behavior with both forward and backscattering elements. Such a two-term H-G function can be written as

$$P(\alpha) = \frac{1+c}{2} \frac{1-b^2}{\left(1 - 2b\cos\alpha + b^2\right)^{3/2}} + \frac{1-c}{2} \frac{1-b^2}{\left(1 + 2b\cos\alpha + b^2\right)^{3/2}} \qquad 4.23$$

In this function, the first term is the backscattering lobe and the second the forward scattering lobe, and c determines which dominates; $-1 \leq c < 0$ is forward scattering, and $0 < c \leq 1$ is backscattering. The width of the lobe is the same for both terms and is determined by b. Figure 4.7 also shows a backscattering example of this function with $b = 0.3$ and $c = 0.2$. The asymmetry parameter of this function is $\xi = -bc$ (Hapke, 2012). There is also a less commonly used three-term version in which the lobe widths in the two terms are allowed to be different.

4.4.7 The L-Shaped Henyey-Greenstein Plot

The phase functions for objects with size parameters $X \ll 1$ (Rayleigh) are mathematically tractable, as are perfect spheres of size $X \sim 1$ (Mie theory) and larger (geometric optics). But particles in most planetary regoliths do not fit

4.4 Particle Phase Function   111

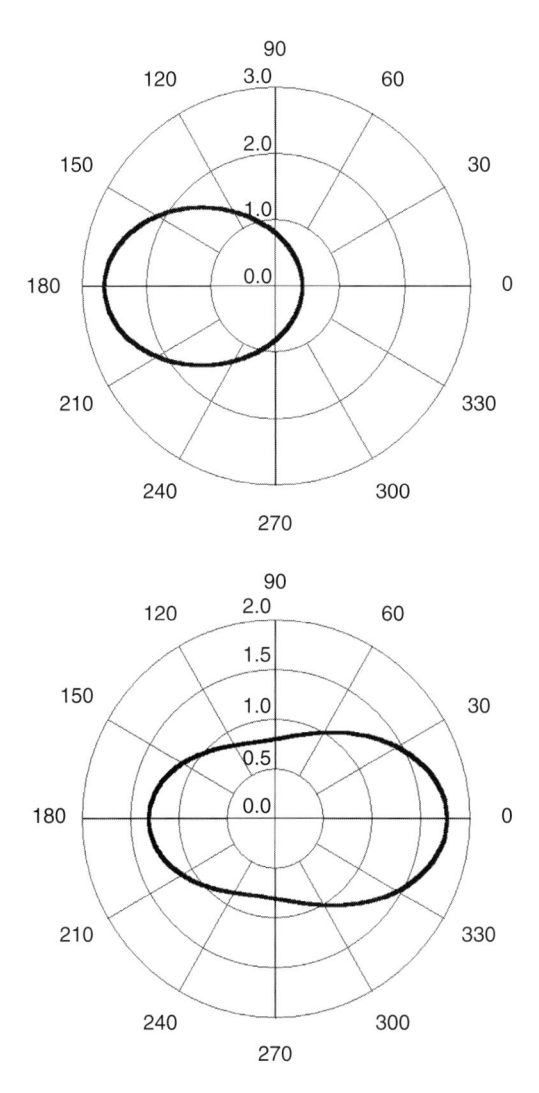

Figure 4.7. H-G phase function plots. Top figure shows a single-term H-G function with $\zeta = 0.3$ (forward scattering). Bottom figure shows a two-term H-G function with $b = 0.3$ and $c = 0.2$ (backscattering).
Credit: Michael K. Shepard.

these criteria; they are on the order of the wavelength in size or larger, but they are roughened, often agglomerates, and full of internal defects. There is no mathematical theory to determine their phase function, so we fall back on smooth empirical phase functions, like the H-G functions.

In 1995, McGuire and Hapke (1995) reported on a series of experiments to measure the phase functions of particulates ($X \gg 1$) that might be more typical of regoliths. They created a series of large particles, each on the order of a centimeter in size. They were composed of glass, colored resin, and metal. Some were left smooth and some were roughened externally. A few of the glass spheres were translucent and full of internal scatterers. Agglutinated particles were created by fusing small blobs of resin into a larger particle. And some particles were created with internal scatterers by casting resin and titanium dioxide grit in a mold. In all, 43 different combinations of particle shape, roughness, and internal scatterer concentration were tested. The amount of light scattered by these particles over a wide range of phase angles, i.e. their particle phase function, was measured using a photometric goniometer (see Chapter 7), and was fit using a two-term Henyey-Greenstein function.

Figure 4.8 is a plot of the region where most particles fell, and a few trends become apparent. Recall from Equation 4.23 that b determines the width of the

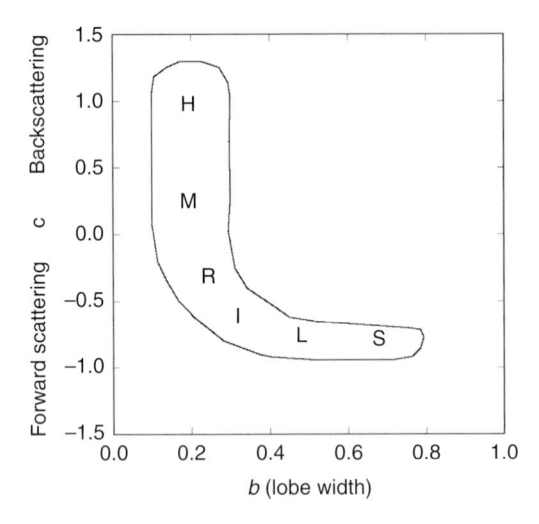

Figure 4.8. Region of the Henyey–Greenstein phase function space occupied by different types of scatterers according to the work of McGuire and Hapke (1995). H = high density of internal scatterers; M = medium density; L = low density; I = irregular particles; R = rough surface dielectrics; S = smooth, clear, and spherical particles.
Credit: Michael K. Shepard.

scattering lobe, and c is the parameter that determines whether back or forward scattering is dominant. Perhaps not unexpectedly, particles with roughened surfaces and large numbers of internal scatterers are more backscattering, and transparent particles are more forward scattering. The width of the lobe (b) falls within a fairly uniform range for most particulate types, and becomes narrow and strong in the forward direction only for the most transparent particles. This plot is referred to by McGuire and Hapke as an "L-shaped" region, but one will find it referred to as "J-shaped" in other publications. One should also be aware (and careful!) that there are different versions of the two-term H-G function (Eq. 4.23) in the literature; one may find signs reversed or the weighting factors (c) scaled differently.

4.5 Scattering by Ensembles of Particles and Radiative Transfer

4.5.1 The Smallest Units of Scattering

In many of the physically based scattering models we will examine, it is assumed that the particles that make up a planetary "surface" are the units of scattering, that they are large compared with the incident wavelength, and that they scatter independently of one another. One can imagine developing a computer simulation of a regolith consisting of millions of simulated particles arranged in a simulated regolith, introducing a light beam from some direction, computing the scattering behavior of each particle, and then summing the contributions of all particles to estimate the overall light scattering properties in any given direction. This type of model may work well for planets shrouded by a thick atmosphere; here the clouds may be composed of dust or droplets that can be modeled as uniform spheres that obey Mie theory or geometric optics and are sufficiently separated so that the assumption of independent scattering is valid. But for the surfaces of most terrestrial planets, moons, and small bodies, these assumptions are subject to challenge.

Regolith returned from the Moon shows particulates with a wide variety of sizes, including a significant fraction at sub-micron scales that cannot be approximated by geometric optics. Many larger particles are translucent and filled with cracks, voids, and inclusions, each of which will scatter incident light. And the close proximity of particles in a regolith make it likely that scattering between adjacent particles is coupled in a non-linear way – we cannot simply add the scattering contributions of each scatterer. All of these greatly complicate the problem.

Even given this depressing pronouncement, there are still reasonable approximations that can be made so that the scattering problem is both tractable and capable of first-order calculations that match what is observed. The most widely used is the Equation of Radiative Transfer.

4.5.2 Radiative Transfer

Although of general use for a variety of physical problems, radiative transfer was originally developed to understand the transport of energy from its origin, deep within a star, to its outer atmosphere and eventual radiation into space (Chandrasekhar, 1960). Subsets of that work relevant to this book include understanding the propagation of light through atmospheres and planetary regolith.

Radiative transfer is very simple in principle, but complex in solution. We ignore individual particles and think of the scattering media as a more-or-less uniform material with gross scattering properties that are a volume-average of the particles that make it up. For any given volume (assumed to be much larger than individual particles), one must keep track of the energy going in, the energy scattered within, the energy absorbed within, the energy within that may be produced there or scattered there from somewhere else, and the energy going out (Figure 4.9). In its most simplified form

$$dL(x) = -\kappa L \, dx + j \, dx \qquad 4.24$$

Here, dL is the change in radiance caused by transmission of energy (light) through a volume over a distance x, with losses, κ, due to absorption and scattering (aka **extinction**), and gains, j, due to an internal source of energy (like thermal emission) and light scattered into the volume from outside (which count as losses for those volumes). In practice, both κ and j can be functions of position, direction of the radiation, wavelength of the radiation, and other physical parameters.

For planetary regolith and atmospheres, it is generally safe (at optical wavelengths) to assume there is no source or energy being produced within a volume. Thus one only need keep track of what goes in, what of that is scattered and absorbed, what is scattered in from outside, and what leaves. If, for the moment, we ignore all sources (j), including light scattered into the volume from outside, Equation 4.24 simplifies to

$$dL = -\kappa L \, dx \qquad 4.25$$

which upon integration over some distance, x, is equivalent to

$$L(x) = L_0 e^{-\kappa x} \qquad 4.26$$

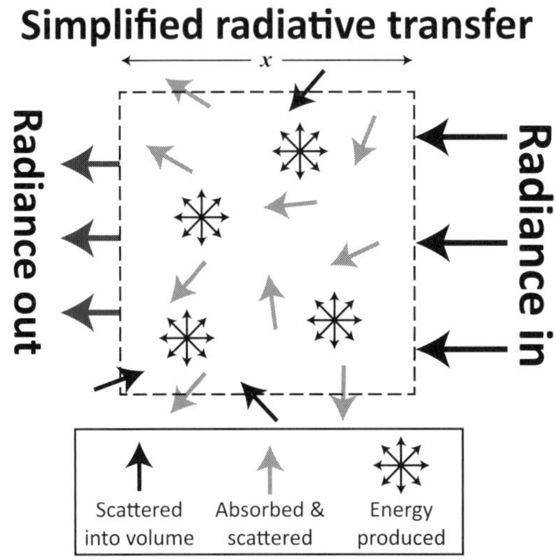

4.5 Scattering and Radiative Transfer 115

Simplified radiative transfer

Figure 4.9. Sketch illustrating the major components of a radiative transfer calculation: the energy entering a volume from a given direction, the fraction of that energy absorbed within or scattered out of the volume, energy scattered from outside into the volume, energy produced within the volume (negligible for most optical applications), and the energy leaving the volume.
Credit: Michael K. Shepard.

In this equation, L_0 is the radiance going in and $L(x)$ is the radiance exiting after passing through a volume of thickness x. This equation is also known as **Beer's Law** (Chapter 1), and κx is referred to as the **optical depth**, $\tau = \kappa x$. If the thickness $x = 1/\kappa$, then $\tau = 1$; we refer to this thickness as one optical depth.

When calculating what happens to an incident light beam as it enters a scattering medium, it is convenient to separate the effects of a single-scattering event and those of multiple scattering. As you might expect, the single-scattering interactions are much simpler to solve, while the multiple-scattering interactions are so complex that analytical approximations or numerical methods are the only viable solutions.

In the next section, we will show how single scattering within the context of radiative transfer leads to one of the most commonly used scattering models, the Lommel–Seeliger model. In the section following that, we briefly look at the multiple-scattering problem.

4.5.3 Single Scattering and Lommel–Seeliger Law

A collimated beam of light of some irradiance, E, shines on a surface regolith which we treat as a homogeneous scattering volume (Figure 4.10). When light

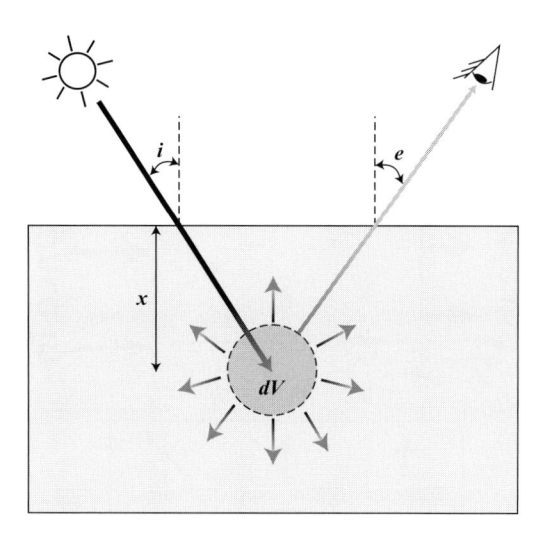

Figure 4.10. Sketch of the factors considered in Lommel–Seeliger scattering. Light enters from a given direction into a small volume, and is scattered. The only light accounted for is that scattered once from the given volume directly into the detector. All other scattered light is ignored.
Credit: Michael K. Shepard.

enters the surface, it encounters individual units (like particles of regolith) where it is either absorbed or scattered. We quantify the likelihood of either event with the **single-scattering albedo**, w (also sometimes as lower case Greek omega, ω) the ratio of the amount of light scattered to the amount of light both absorbed and scattered:

$$w = \frac{Q_{scattered}}{Q_{scattered} + Q_{absorbed}} \quad 4.27$$

Here Q is a kind of probability that light will be affected in a particular way; it is also referred to as the **cross-section** for scattering or absorption, and its units are in terms of area. One may think of a particle as being made of two targets of different areas – if a photon of light hits one target it is absorbed, if it hits the other it is scattered. The likelihood of either event depends upon the area of the target. Surfaces made of transparent materials are more likely to be scattered (high $Q_{scattered}$) and have single-scattering albedos that approach $w = 1$; those made of highly absorbing material (high $Q_{absorbed}$) approach $w = 0$.

Because we are using the methods of radiative transfer and treating the surface as a volume – not as individual particles – the single-scattering albedo defined here is a property of the volume, and is sometimes referred to more accurately as the **volumetric single-scattering albedo**.

4.5 Scattering and Radiative Transfer

The fraction of the incident irradiance that reaches a distance, x, into the regolith is given by Beer's Law (Eq. 4.26)

$$E(x,\mu_0) = Ee^{-\kappa x/\mu_0}$$

or equivalently,

$$E(\tau,\mu_0) = Ee^{-\tau/\mu_0} \qquad 4.28$$

where we have used the definition of optical depth. Note that we have divided by $\cos i$ (μ_0) to take into account the longer path length when the incidence is not normal to the surface.

The Lommel–Seeliger (L-S) scattering model makes two simplifying assumptions. First, if a photon is scattered, it has no preferred direction; this is isotropic scattering. Second, after a photon of light encounters a particle, its subsequent behavior can be ignored. If absorbed, it is lost forever – the energy of the photon is converted to heat (assuming no fluorescence or similar behavior). If it is scattered, it is only scattered once. Even though a real photon may be scattered many times by different particles and eventually emerge back from the surface to be measured, this model counts only the photons that are scattered once.

Given these simplifications, especially the latter one, the Lommel–Seeliger model is only a realistic approximation for dark surfaces. For example, consider two media, one made of particles with $w = 0.9$ and the other with $w = 0.1$. If a photon encounters the bright particle, its chance of scattering once is 90% and its odds of scattering a second time are 0.9^2 or 81%. This is not negligible. However, for the particle with $w = 0.1$, the odds of scattering a first time are 10% and a second time are 1%, which can safely be ignored.

Once a scattering event occurs at some optical depth with likelihood w, the radiance in every direction (4π steradians) is

$$L(\tau,\mu_0,\mu) = \frac{wE}{\mu 4\pi} e^{-\tau/\mu_0} \qquad 4.29$$

Note that the right half is divided by $\cos e$ (μ) because the definition of radiance (Eq. 2.18) requires that we account for the foreshortening due to observing at an angle e. The probability that the resulting radiance will make it back out of the regolith along a direction angle e into a sensor is also given by Beer's Law. Thus the radiance at the sensor contributed by any scattering event within a tiny volume of thickness $d\tau$ at depth τ is given by

$$dL(\tau,\mu_0,\mu) = \frac{wE}{\mu 4\pi} e^{-\tau/\mu_0} e^{-\tau/\mu} d\tau \qquad 4.30$$

118   The Physical Basis of Photometric Scattering Models

To find the total radiance at the sensor, we integrate over all optical depths:

$$L(\mu_0,\mu) = \frac{wE}{\mu 4\pi} \int_0^\infty e^{-\tau\left(\frac{1}{\mu_0}+\frac{1}{\mu}\right)} d\tau$$

$$L(\mu_0,\mu) = \frac{wE}{\mu 4\pi} \frac{\mu_0\mu}{\mu_0+\mu} = \frac{wE}{4\pi}\frac{\mu_0}{\mu_0+\mu} \qquad 4.31$$

The functional form of the L-S model is therefore (dividing by the irradiance)

$$r = \frac{w}{4\pi}\frac{\mu_0}{\mu+\mu_0} \qquad 4.32$$

where r is the bidirectional reflectance of a surface (Chapter 5).

4.5.4 Multiple Scattering and the H-functions

Multiple scattering greatly complicates the solution to the problem of finding the reflectance of light from or its transmission through a medium. The problem is that every scattering event out of a small volume in the media becomes a source for some adjacent volume, and these must be counted if our solution is to be accurate. This additional source of illumination is shown in Figure 4.11 as arrows pointing into the scattering volume.

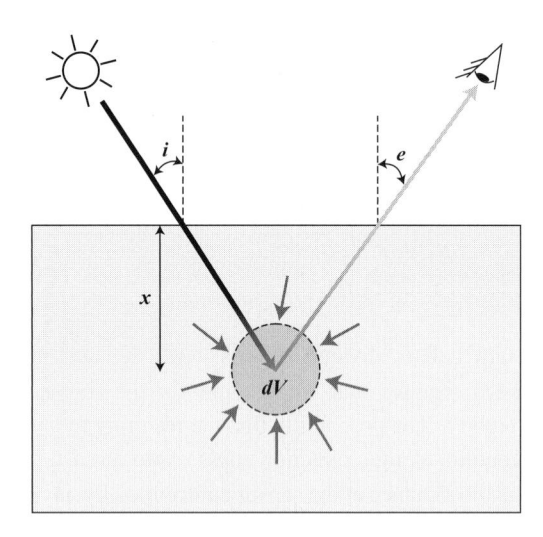

Figure 4.11. For a more realistic treatment of scattering using radiative transfer, the Lommel–Seeliger model must also account for light scattered from elsewhere into the volume and then to the detector (multiple scattering).
Credit: Michael K. Shepard.

4.5 Scattering and Radiative Transfer

In the Lommel–Seeliger solution provided earlier, this was ignored and justified by limiting it to media with small single-scattering albedos. But as the single-scattering albedo increases, this approximation becomes increasingly worse.

The problem of estimating the multiple-scattering component has been the source of much work and, although no exact analytic solution has been found, there are a number of approximations which work well for simple cases. Those solutions are beyond the scope of this book, but the interested reader is encouraged to consult the more advanced papers and books on the topic (Chandrasekhar, 1960; Kortum, 1969; Hapke, 2012). Here, we will only give the final solutions and place them in context.

The simplest case of multiple scattering assumes that the scattering is isotropic, i.e. there is no preferred scattering direction. In the buildup to a solution of this problem, Hapke (1993) introduces the concept of diffuse reflectance, r_0. This parameter is the ratio of the total amount of light re-emitted (scattered) from a semi-infinite half-space to the incident illumination. The illumination is uniform and isotropic. All of the scattering takes place at the surface boundary – there is no source penetration into the medium. Hapke notes that this is not realistic, but that the concept is useful because it appears in solutions to the full radiative transfer problem.

$$r_0 = \frac{1-\gamma}{1+\gamma} \qquad 4.33$$

where

$$\gamma = \sqrt{1-w} \qquad 4.34$$

Hapke (2012) also notes that the diffuse reflectance is analogous to the single-scattering albedo; w is the amount of light scattered by a particle and r_0 is the amount of light scattered out of an isotropically scattering medium.

A related but more realistic parameter than the diffuse reflectance is the *bi-hemispherical reflectance*. It is defined the same as the diffuse reflectance except that the illumination can penetrate into the medium before scattering. We will not pursue it further except to note that it is the same as the Bond albedo (Hapke, 2012), a parameter to be defined in the next chapter.

If a collimated light source is incident upon a medium of isotropic scattering, Hapke (2012) shows that the bidirectional reflectance is given by

$$r = \frac{w}{4\pi} \frac{\mu_0}{\mu + \mu_0} H(\mu) H(\mu_0) \qquad 4.35$$

where $H(x)$ is given by

$$H(\mu) = 1 + \frac{w}{2}\mu H(\mu) \int_0^1 \frac{H(x)}{\mu + x} dx \qquad 4.36$$

Because $H(\mu)$ is recursive, exact solutions require numerical methods. However, there are several good approximations.

According to Hapke (2012)

$$H(\mu) = \frac{1 + 2\mu}{1 + 2\gamma\mu} \qquad 4.37$$

is good to within 4% of the exact solution. Hapke (2012) also provides a more complex equation that is accurate to within 1%:

$$H(\mu) = \left\{ 1 - [1 - \gamma]\mu \left[r_0 + \left(1 - \frac{1}{2}r_0 - r_0\mu\right) \ln\left(\frac{1+\mu}{\mu}\right) \right] \right\}^{-1} \qquad 4.38$$

With the ubiquity of the desktop computer, it is also easy to build a lookup table of exact solutions for H generated by numerical methods.

In the more complex scattering models considered in Chapter 5 (Hapke, Lumme–Bowell), multiple scattering is included. However, most of these models assume the scattering function to be isotropic ($P(\alpha) = 0$; $\xi = 0$) for the multiply scattered component while allowing the singly scattered component to have an anisotropic phase function. The justification for this is that multiple-scattering randomizes the direction of scattering sufficiently for the isotropic approximation to be realistic. This solution is often referred to as the **isotropic multiple-scattering approximation** (IMSA). An analytical method of including anisotropic scattering into the multiple-scattering terms was introduced by Hapke (2002).

4.5.5 Limitations of Radiative Transfer

The equations of radiative transfer are generally sufficient to accurately model scattering within planetary atmospheres. The scattering centers, gas or dust, are far enough apart to meet the assumption of independent and isolated scattering elements. However, when applying the equations to planetary regolith, these assumptions are often violated.

There are two major phenomena that cannot be handled by radiative transfer: the rapid brightening observed in regolith as one approaches $\alpha = 0$ (opposition surge), and surface boundaries that are roughened instead of plane. Scattering models based on radiative transfer like those of Hapke (2012) and Lumme and Bowell (1981) include modifications to address these two important features (Chapter 5).

4.6 Opposition Surge

4.6.1 Definition and Early Observations

The **opposition surge** is the observation that a planet brightens at an accelerating rate as it approaches opposition, i.e. phase angles approach 0°. Most authors use the term "non-linear" brightening, but this is only technically correct when the magnitude of a planet (radiance on a logarithmic scale) is plotted versus its observed phase angle. The *magnitude* is approximately linear from $\alpha = 10°$ to $30°$ or so (the radiance would be non-linear), which is the maximum phase angle for most objects when observed from Earth, and becomes non-linear for $\alpha < 10°$.

The phenomenon was noted in the rapid brightening of Saturn's rings near opposition by the German astronomer **Hugo von Seeliger** (1849–1924) (von Seeliger, 1887; Hapke, 2012). Even though the geometry of the Sun–Earth–Moon only allows us to observe the Moon down to a phase angle of ~2°, early phase curves of the Moon also displayed the phenomenon (e.g. Russell, 1916a; Barabashev, 1922; Markov, 1924; Opik, 1924). The full Moon is roughly ten times as bright as a first or third quarter Moon and the non-linear surge is evident in the plot of the Moon's whole-disk phase curve (Figure 1.5 in Chapter 1; also see Lane and Irvine, 1973). Gehrels (1956) was the first to observe the rapid surge in an asteroid (20 Massalia) and appears to have been the first to use the term **opposition effect**, which is still in use today. Opposition effects have since been observed in the phase curves of most of the solid planetary objects observed at the proper geometry (Figures 4.12 and 4.13).

4.6.2 Empirical Models

Before the development of physically motivated opposition effect treatments, empirical models were used to probe the behavior and explore its possible dependence on properties such as albedo and color. Two aspects of the opposition surge must be treated in all models: the amplitude of the surge and its angular width. In empirical models, these are often treated with simple mathematical expressions, such as an exponential function that has both an amplitude coefficient B_0 and an exponential scale rate C_0 that defines the phase angle-dependent shape of the surge.

$$r(i, e, \alpha) = B(\alpha) F(i, e, \alpha) = B_0 \exp(-C_0 \alpha) F(i, e, \alpha) \quad 4.39$$

where $r(\alpha)$ is the reflectance of the surface, $F(i, e, \alpha)$ is a function that models the reflectance behavior of the surface in the absence of an opposition effect, and $B(\alpha)$ is a function that models the shape of the opposition effect.

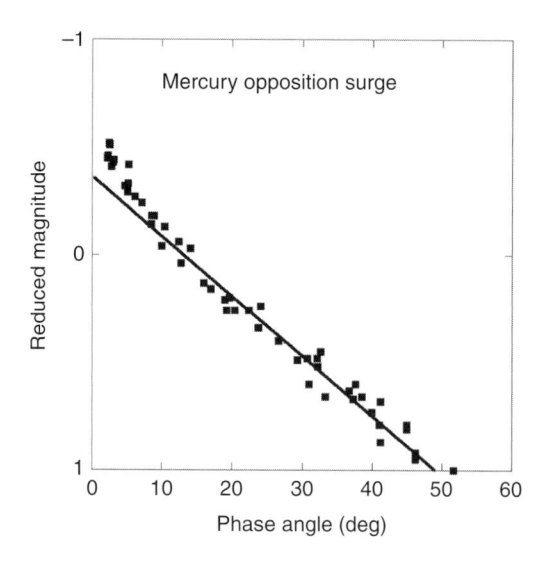

Figure 4.12. A partial phase curve of Mercury. The line shows the behavior from phase angles of ~10° to 100° (not all shown). The opposition surge is the non-linear brightening for a < 10°.
Credit: Data from Mallama et al. (2002).

Figure 4.13. The opposition surge is illustrated by photos taken by Apollo 17 astronaut Jack Schmitt as part of a panoramic sequence. The photo on the left shows phases near 0°, while the photo on the right shows phase angles closer to 180° (the Sun is out of the frame just to the left). The brightening around the astronaut's shadow is due to the SHOE.
Credit: Adapted from photos AS17-147–22495HR and AS17-147–22511HR. Courtesy: NASA. www.hq.nasa.gov/alsj/a17/images17.html

4.6.3 Shadow Hiding

Although von Seeliger (1895) is often given credit as the first to propose shadow hiding in his investigation of the phenomenon in observations of Saturn's rings, it was also proposed by Zöllner (1865) to explain the non-linear brightening of the full moon. The shadow hiding mechanism is based on the phenomenon that every roughened or particulate surface casts shadows. They can be of two types; shadows cast by the light source incident at an angle other than directly overhead, also known as **illumination shadows** ($i > 0°$), and "shadows" that block an observer's view of things behind it, or **viewing shadows**, caused by the observer viewing the surface at an angle other than directly overhead ($e > 0°$). In most illumination and viewing conditions, both types of shadows are present. However, if the observer and light source are in a direct line, i.e. the object is at opposition ($\alpha = 0°$), then both types of shadows hide each other and the object is maximally illuminated as perceived by the observer. The transition to this condition changes very rapidly near opposition, thus giving rise to the accelerated brightening observed (Figure 4.14). In most photometric models, the shadows cast by topographic roughness are considered separately from those cast by a porous particulate surface because, even for very rough surfaces, they are insufficient to explain the observed phenomenon. We will consider the shadowing caused by topography later in this chapter.

For particulate surfaces, it is expected that there is considerable pore space between particles. In this conception of the **shadow hiding opposition effect**

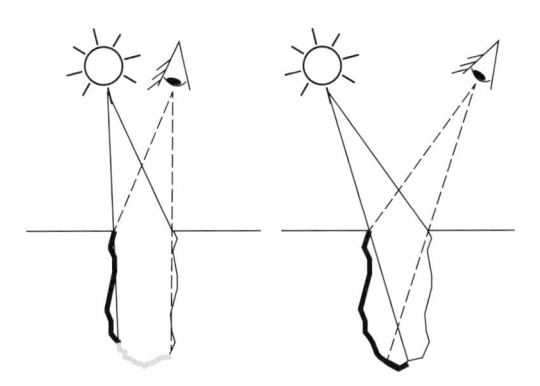

Figure 4.14. A sketch illustrating the shadow hiding opposition surge. When source and detector are at small phase angles, the detector sees mostly illuminated portions of the surface (shown as the thicker light gray line). At higher phase angles, an increasing portion of the view includes shadows (shown as the thicker dark line).
Credit: Michael K. Shepard.

(**SHOE**), some light will penetrate deeper into the regolith without scattering via pore spaces. If the incident and viewing angles differ, the light traveling through one set of "tunnels" is unlikely to emerge into the viewing direction without first encountering another particle and being scattered or absorbed. If, however, the incident and viewing directions co-align, the same path is available for the light to enter and exit the regolith (Figure 4.14). As the incident and viewing directions get closer (α approaches $0°$), the shadows disappear quickly, giving rise to the sudden brightening.

Hapke (2012) provides a simple estimate of the angular width of the SHOE. In a particulate medium, assume the average particle radius is s. If we characterize the average distance between extinction events (scattering or absorption) in that medium as x_{ext}, then this will be also the average length of a particle's shadow. The base of the shadow will not be visible to an observer at $\alpha = 0°$, but as α increases, more of the shadowed floor will be seen. As a rough approximation, half of the shadow will be visible when

$$\alpha \cong \frac{s}{x_{ext}} \qquad 4.40$$

where α is in radians. This is a reasonable estimate for the half-width half-maximum (HWHM) of the SHOE. Although it is often assumed that SHOE exhibits no wavelength dependence, the extinction length, x_{ext}, depends on the albedo and transparency of the constituent regolith grains which may change with wavelength.

If the particles are far apart compared with their size, the angular width effectively goes to zero. This is why we observe SHOE in particulate media, but not in dispersed clouds of dust or water droplets. Since the objects making up Saturn's rings are dispersed, it is likely that von Seeliger was wrong about the cause of the opposition surge in Saturn's rings. The rings are also mostly bright ice, which adds another wrinkle.

Shadow hiding is a single-scattering phenomenon and relies upon the contrast of shadows with a media. The brighter the surface (larger w), the more multiple scattering will tend to wash out shadows and reduce the magnitude of the effect. For many years, the expectation that the opposition surge only applied to darker objects in the solar system was fulfilled (with the exception of Saturn's rings), and shadow hiding was the only mechanism considered viable. The additional discovery of very narrow but significant opposition surges from other bright objects, including some asteroids (Harris et al., 1989; Mishchenko and Dlugach, 1993) and icy moons (Hapke and Blewett, 1991; Mishchenko, 1992), required a revision of this thinking and the search for a second opposition mechanism.

4.6.4 Coherent Backscattering

This second mechanism for the opposition surge is called the **coherent backscattering opposition effect** (**CBOE**). It was proposed by a number of planetary scientists including Shkuratov (1989), Muinonen (1990), Hapke (1990), and Mishchenko (1992). The idea is that for any given path in which a wave of light is scattered two or more times before returning to the observer, the wave can travel either in an identical forward or backward direction along the same path (Figure 4.15). Because these waves travel exactly the same distance, only differing in the direction of their route, they combine coherently at $\alpha = 0°$, potentially doubling the intensity of the reflection that would otherwise be expected.

The coherence of the forward and backward traveling waves falls off quickly as one leaves the exact opposition direction, giving rise to narrow opposition spike. The width of the effect is a function of the distance between the first and last particles in the chain, x. For any given phase angle, the forward and backward waves travel paths that differ by distance z, given by

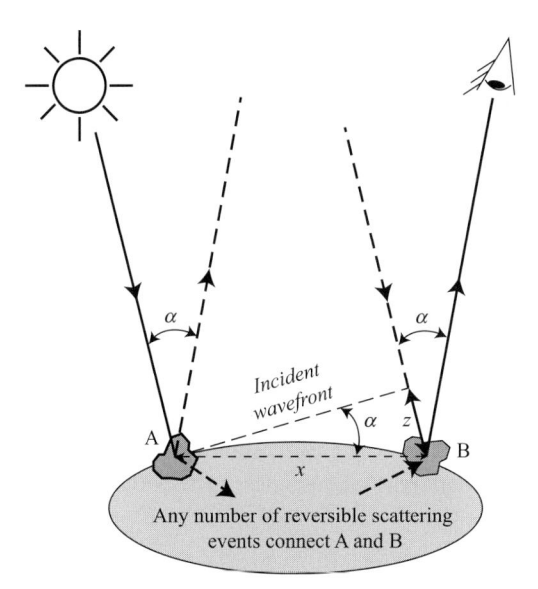

$$z = x \sin \alpha \qquad 4.41$$

Figure 4.15. Sketch illustrating the mechanism responsible for the coherent backscatter opposition effect. Light rays travel the exact same path but in reversed directions and constructively cohere. The coherence decreases as the phase angle increases.
Credit: Michael K. Shepard.

and therefore differ in phase, $\Delta\varphi$, by

$$\Delta\varphi = \frac{2\pi z}{\lambda} = \frac{2\pi x \sin\alpha}{\lambda} \qquad 4.42$$

We can estimate the width of the CBOE by rewriting this to solve for phase angle and note that $\sin\alpha \sim \alpha$ for small angles in radians:

$$\alpha = \frac{\lambda \Delta\varphi}{2\pi x} \qquad 4.43$$

Waves π radians out of phase will destructively interfere, so a rough estimate of the half-width half-maximum (HWHM) of the CBOE is

$$\alpha = \frac{\lambda}{4x} \qquad 4.44$$

The physics of this mechanism suggests that the CBOE will be most prominent in bright surfaces where multiple scattering is more likely. Equation 4.44 also indicates that the width of the opposition surge varies in proportion to the wavelength of incident light, and in inverse proportion to the optical path length x. For a given material, x is a function of particle density, size, and scattering and absorption properties. Materials with a low absorption coefficient (high single-scattering albedo, w) should have a significant contribution of CBOE from large distances, x, giving a strong but narrow surge. Surfaces composed of materials with a high absorption (low w) should have fewer multiply scattered waves, and of those, the average distance between scattering events should be smaller. Therefore, in dark objects, any coherent backscattering should be weak and wide, and difficult to discern from the SHOE.

In nature, the observed opposition effects of regolith-covered surfaces is probably a superposed combination of SHOE and CBOE. At the extreme limits, we expect that SHOE should be absent in highest-albedo surfaces, and for CBOE to be absent in the lowest-albedo materials. However, a reliable, simple expression for predicting the relative contribution of each mechanism has yet to be found. A narrow CBOE contribution exists even in soils composed of relatively dark regolith particles, such as those in lunar soils (Hapke et al., 1993, 1998).

4.7 Surface Roughness

Scattering models, either empirical or based on the principles of radiative transfer, usually assume a smooth surface interface. Roughening the surface violates this assumption and, depending on the amount of roughening, can become an important factor.

Some of the first models of diffuse scattering made the assumption that geometric optics was the operative scattering law, but that it was modified by the deviations of the surface – by surface roughness. The surface was conceived of as consisting of a myriad of tiny facets, each individually smooth and large enough so that Fresnel reflection could be assumed. To be considered "smooth," the vertical displacement of each facet must deviate less than $\lambda/4$ from the plane. The earliest record of the surface facet hypothesis appears to be by Bouguer in the late 1750s (Berry, 1923; Kortum, 1969). Berry (1923) shows that, depending on the statistical distribution of facets assumed, this type of model can generate a diffuse scattering pattern that is pseudo-Lambertian.

4.7.1 Characterization of Surface Roughness

4.7.1.1 Stationary Random Surfaces

The assumption of a plane surface with some statistical variation about the mean is still a staple of many surface roughness models. In mathematical terms, this type of surface is a **stationary random process**, meaning characteristics like the **root-mean-square (RMS) height** (square root of the height variance) do not change over distance or scale. White noise is an example. Such a surface can be characterized by a mean height (typically normalized to zero) and a model distribution of heights about the mean, often assumed to be Gaussian. With a Gaussian distribution model, it is common to report the standard deviation or variance of heights (or equivalently slopes) about the mean as a measure of roughness.

Many of these stationary surface roughness models also utilize a parameter called the **autocorrelation length**, which is found from the **autocorrelation function**. The autocorrelation function, also referred to as the **lagged correlation** or **correlogram**, measures the change in correlation between points as a function of their proximity. To illustrate, consider any surface. Pick a point on that surface and ask – how well can we use the knowledge of that point's height or slope to predict the value of an adjacent point? If our surface is a flat plane, then our knowledge of any given point's height will exactly tell us the values of adjacent points – they are perfectly correlated. If our surface is white noise, however, then the height or slope at any point tells us nothing about the values of adjacent points – they are uncorrelated. Most surfaces are in between these extremes and the correlation drops as one moves away from a point.

In practice, surfaces are treated as arrays of discretely measured heights with uniform spacing and we can find the autocorrelation between points separated by a distance or **lag**, x, along any given profile of N points along such a surface using

$$f(x) = \frac{\sum_{i=1}^{N-1}(h_i - \bar{h})(h_{i+x} - \bar{h})}{\sum_{i=1}^{N-1}(h_i - \bar{h})^2} = \frac{\sum_{i=1}^{N-1}(h_i - \bar{h})(h_{i+x} - \bar{h})}{\sigma_h^2} \qquad 4.45$$

where h_i is the height of the point at i along the profile, $f(x)$ is the autocorrelation function at a lag x, and σ_h^2 is the variance of heights about the mean. Inspection of Equation 4.45 shows that, when $x = 0$, $f(x) = 1$, so the function is normalized. When $f(x) = 0$, there is no correlation. It is possible for points to be anti-correlated at some lags – for example if there is periodicity in the profile – so in general, $-1 \leq f(x) \leq 1$. Often, the autocorrelation function is modeled as an exponential decay function, and the autocorrelation length is then usually defined to be the distance x for which $f(x) = 1/e$.

This type of stationary model is mathematically tractable and convenient, and will be found associated with many scattering models, including Hapke and Lumme–Bowell (Chapter 5). However, it is not a realistic model for most natural surfaces. What it lacks is a realistic handling of scale.

4.7.1.2 Fractal or Self-Affine Surfaces

Natural surfaces of planets are rarely stationary, except in very restricted cases. To show this, consider a stationary random surface profile 10 m in length. The surface height is measured every centimeter and normalized to a zero mean, and the RMS height about that mean for this example profile is 5 cm. If we now took a 100 m profile of the same surface, the RMS height about the mean would again be 5 cm. But this is not what we see in nature. A 10 m profile of a lava flow might also have an RMS height of 5 cm. But a 100 m long profile over the same surface would have a significantly larger RMS height; using behavior typical of many surfaces, we would expect the RMS height to be ~15 cm. This behavior is generically referred to as **fractal**, and more precisely as **self-affine**.

The statistical parameters of a self-affine profile or surface will change as the scale of measurement changes. A simple relationship that appears to hold for the most natural surfaces is

$$\sigma_h(L) = \sigma_h(L_0)\left(\frac{L}{L_0}\right)^H \qquad 4.46$$

where $\sigma_h(L)$ is the RMS height (aka standard deviation of heights) for a profile of length L, L_0 is a standard or unit length, and H is a parameter called the

4.7 Surface Roughness

Hurst exponent ($0 \leq H \leq 1$) (Shepard et al., 2001). If $H = 0$, the RMS height is independent of profile length – the profile is stationary. If $H = 1$, the profile is self-similar so that the RMS height scales proportionally with the size of the profile over which it is measured. Most natural surfaces are well approximated when $H = 0.5$; this is referred to as **Brownian** behavior (Shepard et al., 2001). Thus, in our earlier example, if a 10 m long profile has a measured RMS height of 5 cm, we would expect a 100 m profile to have an RMS height of $10^{0.5} \sim 3$ times higher, or 15 cm.

A similar relationship holds for the other commonly measured parameter, the RMS slope, s_{RMS}. The slope between two points on a profile is their difference in height divided by their distance apart, Δx. The RMS slope is therefore

$$s_{RMS}(\Delta x) = \left[\frac{1}{N-1} \sum_{i=1}^{N} \left(\frac{h_i - h_{i+\Delta x}}{\Delta x} \right)^2 \right]^{\frac{1}{2}} \qquad 4.47$$

The behavior of RMS slope with scale on a self-affine profile will be well approximated by

$$s_{RMS}(\Delta x) = s_{RMS}(\Delta x_0) \left(\frac{\Delta x}{\Delta x_0} \right)^{H-1} \qquad 4.48$$

(Shepard et al., 2001) where Δx_0 is a standard or unit distance. If a profile is self-similar, $H = 1$, we expect its RMS slope will be the same at all scales; if stationary, $H = 0$, then RMS slope will decrease linearly with scale.

Despite their more realistic treatment of natural terrain, self-affine models have not yet been utilized in most scattering models.

4.7.2 Shadowing Models

In virtually every modern photometric model that purports to account for roughness, surfaces are still thought of as consisting of small facets that deviate from horizontal in a statistically quantifiable way. Each facet is then presumed to scatter according to the same scattering function. While the previously discussed models of Bouguer and Berry assumed each facet obeyed geometric optics and Fresnel reflection, more modern treatments assume each facet scatters diffusely in accordance with an empirical or radiative transfer model (Hapke, 1984).

The tilted geometry of each facet changes the overall scattering behavior of the surface in two ways. First, individual facets are unlikely to be horizontal, so their local lighting and viewing geometry differ from the global, and each facet will scatter the incident light slightly differently as a result. To a first approximation, one can argue that the scattering from all the facets will tend to average out to the mean of a flat surface. However, if the scattering law deviates significantly from isotropic or if the surface is very rough, this averaging approximation will be poor.

The second, and arguably the more important way that roughness modifies a scattering law is because of shadowing. At any incidence or emission angle other than $0°$, some parts of the surface will be in shade from the light source or the detector or both. One of the significant analytical difficulties in accounting for these is that they are not independent of one another – cast shadows may or may not be hidden from view, depending on the specific geometry.

There have been a number of shadowing models published, but we do not consider them except in the context of their specific photometric models (Chapter 5). However, the interested reader will find a review and comparison of several shadowing models, including one for self-affine surfaces, in Shepard and Campbell (1998).

Finally, it is important that the reader divest themselves of the notion that roughness in a photometric model is equivalent to a physically measured roughness. Shadowing models require facets and shadows, but these concepts break down as one looks to finer and finer scales. Regolith particles are not facets, and while they shade adjacent particles, they also transmit light. At some point, shadows disappear, filled in by transmitted and multiply scattered light. One would expect this to occur sooner (at larger scales) in brighter surfaces, so even a direct comparison of photometric roughness between bright and dark surfaces is likely one of apples and oranges.

5

Planetary Reflectance and Basic Scattering Laws

One hundred years ago, Henry Norris Russell (1877–1957), one of the giants of modern astronomy, wrote two important papers in the field of planetary photometry. The first, *The Stellar Magnitudes of the Sun, Moon, and Planets* (Russell, 1916a) compiled the best data up to that point for the fundamental brightness of the objects in our solar system. At that time, making precise magnitude measurements of stars and planets was a major undertaking, calling for sophisticated instrumentation and analyses. Today, we take these routine measurements for granted.

In the second of his 1916 papers, *On the Albedo of the Planets and Their Satellites* (Russell, 1916b), Russell developed or reiterated much of the theory of planetary photometry. His end goal was to take the brightness (i.e. magnitude) measurements of planets and calculate their albedo – a measure of how much light they reflect – in order to say something about their physical makeup. But this requires a scattering model, a mathematical relationship that tells us the amount of reflected light expected at any given illumination and viewing geometry.

In this chapter, we will examine the most commonly used parameters for describing the reflectance and albedo of planetary surfaces; first for disk-resolved, and later for disk-integrated measurements. As a cautionary note, we will point out that some of these terms are used in different ways, or have different names given to them by different authors. We have tried to note this where possible, but we almost certainly have not caught all instances. Caveat emptor.

We will also look in some detail at the most common scattering models used up until mid-1980s. We have relied heavily on the excellent papers by Minnaert (1941, 1961), Lester et al. (1979), Hapke (1981), Fairbairn (2004, 2005), and Shkuratov et al. (2011). In most cases, we include detailed derivations so the reader can understand the source of the final equations, if desired. Summaries of the definitions and relations are included in Tables 5.1 and 5.2.

Table 5.1 *Disk-Resolved Quantities (Applicable to Flat Surfaces)*

Law	r	r_n	r_h	$f(i,e,\alpha)$
Lambert	$\frac{k}{\pi}\mu_0$	k	k	μ_0
Lommel–Seeliger	$\frac{w}{4\pi}\frac{1}{\mu_0+\mu}$	$\frac{\pi}{8}$	$\frac{w}{2}\left[1-\mu_0 \ln\left(1+\frac{1}{\mu_0}\right)\right]$	$\frac{2\mu_0}{\mu+\mu_0}$
Minnaert	$\frac{k}{\pi}\mu_0^n\mu^{n-1}$	k	$\frac{2k}{n+1}\mu_0^{n-1}$	$\mu_0^n\mu^{n-1}$

Table 5.2 *Disk-Integrated Quantities*

Law	p	q	A	Φ
Lambert	$\frac{2}{3}k$	$\frac{3}{2}$	k	$\frac{1}{\pi}[\sin\alpha + (\pi-\alpha)\cos\alpha]$
Lommel–Seeliger	$\frac{w}{8}$	$\frac{16}{3}(1-\ln 2)$	$w\frac{2}{3}(1-\ln 2)$	$1+\sin\frac{\alpha}{2}\tan\frac{\alpha}{2}\ln\left(\tan\frac{\alpha}{4}\right)$
Area	πk	2	$2\pi k$	$\frac{1+\cos\alpha}{2}$

5.1 Lambertian Reflectance

There are two Lambert cosine "laws" in the literature, named for Johann Lambert. The first is the ***cosine incidence law*** – the power incident on a surface is proportional to the cosine of the incidence angle. The second is the ***cosine emission law***, also known as Lambertian reflectance (see Chapter 1).

Lambertian reflectance is the simplest type of scattering behavior from a surface or sphere, and often the primary standard by which other surfaces are compared. An ideal or perfect Lambertian surface has two important properties. First, it reflects 100% of the incident light back into the upper hemisphere – there is no transmission through or absorption. While not completely achievable in practice, there are a number of manufactured substances, such as barium sulfate paint and Spectralon® (Springsteen, 1989, 1999), which come very close (Figure 5.1). Second, if the light source is held constant, a Lambertian surface appears to have the same brightness (radiance), independent of the viewing direction. Again, this is not realized in practice, but there are a number of materials that come close – Spectralon is one, common white bond paper is another.

There are also Lambertian surfaces which are not "perfect." They are Lambertian in the sense that they appear to be of the same brightness independent of illumination or viewing direction, but do not reflect 100% of the incidence light. One may purchase a number of commercial products that are

Figure 5.1. A commercial Lambertian standard.
Credit: Michael K. Shepard.

of varying shades of gray, sold as Lambertian standards, but with a fixed albedo between 0.0 and 1.0. These are used as calibration targets in applications where it is desired to have a linear brightness scale. Where needed, we will use the constant k to quantify the fraction of light reflected from these less-than-perfect Lambertian surfaces.

Mathematically, the directional reflectance property of a Lambertian surface is

$$L_{Lam}(i, e, \alpha) = L_{Lam}(0, 0, 0) \cos i = L_{Lam}(0, 0, 0)\mu_0 \qquad 5.1$$

Let's consider this for a moment. If a surface is illuminated by a source that is not vertical, the amount of power incident on the surface is reduced by $\cos i$, as shown in the equation. This is the *Lambertian cosine incidence law*. However, the radiance remains the same, regardless of the viewing or emission angle.

Intuitively, one might expect the radiance to have some directional component – perhaps more will be reflected in the direction opposite the source, as a mirror might. However, a Lambertian surface will show no change in the radiance with emission angle, even if the source is well off-axis. This does not mean that a Lambertian surface reflects light equally, or isotropically, in all directions. It does not, although it *appears* to do so. In fact, the amount of power reflected from a Lambertian surface *decreases* at higher emission angles in proportion to $\cos e$. How does this make sense?

When you or a sensor observe a surface at an angle other than straight down ($e \neq 0°$), the amount of area you *think* you see stays the same, but in reality, you are looking at an ever-increasing area because of the projection effect. The increase in illuminated surface area is $dA/\cos e$ and exactly balances the

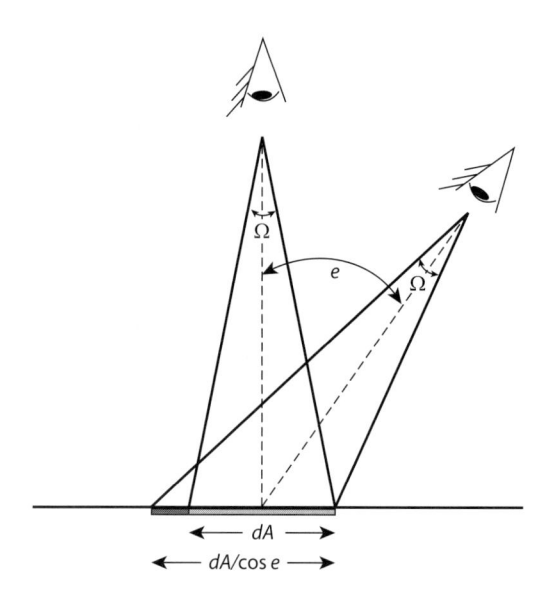

Figure 5.2. An illustration of the Lambert cosine emission law, showing how the apparent area of a scene changes with emission angle.
Credit: Michael K. Shepard.

decrease in the amount of energy reflected per area, so that the overall radiance stays the same (Figure 5.2). This is the *Lambertian cosine emission law*. Given this behavior, what is true is that the *radiance* of a Lambertian surface is isotropic. Lambert did not introduce any theoretical underpinning to explain such behavior, but found that it approximated the scattering behavior of many (especially brighter) diffuse surfaces.

If a Lambertian surface is illuminated by a source, what is the relationship between the incoming irradiance, E, and the outgoing radiance, L? To calculate this, we must sum all the contributions of outgoing radiance (the upper hemisphere) and remember that the energy-in must balance the energy-out. The constant k accounts for any energy absorbed, and equals 1.0 for a perfect Lambertian reflector. For a given area, dA, illuminated by the Sun (solar flux density, F, see Chapter 2), we can write the incoming and outgoing energy as:

$$kF\mu_0 dA = kE\, dA = \int_0^{2\pi} L_{Lam}\, d\Omega \cos e\, dA \qquad 5.2$$

The $\cos e$ is necessary because, for a fixed area on the surface, the solid angle decreases as the viewing angle increases (or alternatively, the actual area viewed increases as viewing angle decreases – the cosine keeps the viewed

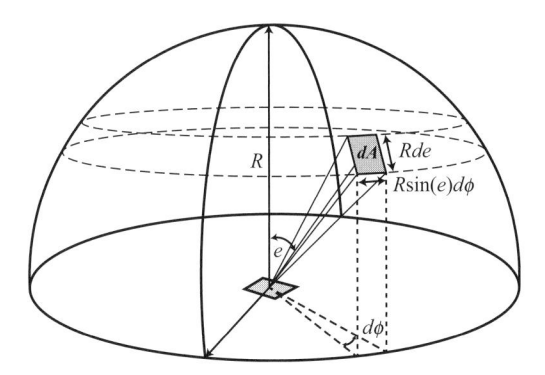

Figure 5.3. Geometry of a small element of a sphere used to find the solid angle. Credit: Michael K. Shepard.

area constant). The areas cancel and we make the substitution for solid angle, $d\Omega = \sin e \, de \, d\phi$ (Figure 5.3) giving

$$kE = \int_0^{2\pi} d\phi \int_0^{\pi/2} L_{Lam} \cos e \sin e \, de = \pi L_{Lam} \qquad 5.3$$

Therefore, the radiance measured from a Lambertian surface is

$$L_{Lam} = \frac{kE}{\pi} = \frac{kF\mu_0}{\pi} \qquad 5.4$$

and if the surface is a perfect Lambertian, $k = 1$ and this reduces to

$$L_{Lam} = \frac{E}{\pi} = \frac{F\mu_0}{\pi} \qquad 5.5$$

5.2 Disk-Resolved Albedos and Measures of Reflectance

5.2.1 Bidirectional Reflectance, *r*

This is the simplest parameter for quantifying reflectance. It is the ratio of the measured radiance from a surface to the collimated irradiance incident upon it (flux density, or power/area perpendicular to the direction of travel).

$$r(i, e, \alpha) = \frac{L(i, e, \alpha)}{F} \qquad 5.6$$

The bidirectional reflectance is thus a function of the incidence and emission angles of the source and sensor, and their position relative to each other quantified by the phase angle, α. Because radiance has units of W m^{-2} sr^{-1} and irradiance has W m^{-2}, the units of bidirectional reflectance are inverse solid angle, sr^{-1}.

For a Lambertian surface

$$r = \frac{k}{\pi}\mu_0 \qquad 5.7$$

For a perfect Lambertian surface, $k = 1$ and $r = \mu_0 \pi^{-1}$.

5.2.2 Bidirectional Reflectance Distribution Function, r_{brdf}

Very similar to the bidirectional reflectance, the bidirectional reflectance distribution function differs only in that it is the ratio of the observed radiance to the incident irradiance. It also has units of sr^{-1}.

$$r_{brdf}(i, e, \alpha) = \frac{L(i, e, \alpha)}{E(i)} = \frac{L(i, e, \alpha)}{\mu_0 F} = \frac{r}{\mu_0} \qquad 5.8$$

For a Lambertian surface of reflectance k

$$r_{brdf} = \frac{k}{\pi} \qquad 5.9$$

For a perfect Lambertian surface, $r_{brdf} = \pi^{-1}$.

5.2.3 Radiance Factor, r_f

In some references, this is also called the *apparent albedo* or *visible albedo* (e.g. Shkuratov et al., 2011). Perhaps the easiest type of reflectance to measure in a laboratory setting, the radiance factor is the ratio of the radiance of a sample at a given illumination and viewing geometry, $L(i, e, \alpha)$, to the radiance of a Lambertian sample, illuminated and viewed from overhead ($i = 0°, e = 0°$).

$$r_f = \frac{L(i, e, \alpha)}{L_{Lam}(0, 0, 0)} = \frac{\pi L(i, e, \alpha)}{F} = \pi r = \pi\mu_0 r_{brdf} \qquad 5.10$$

The middle term in the preceding equation places the radiance factor in terms of the flux density, and the latter terms show the relationship to the bidirectional reflectance and bidirectional reflectance distribution function.[1] In most

---

[1] As noted in a previous footnote (Chapter 2), some write the incident flux density as πF where F is the *radiance* from a perfect Lambertian surface illuminated and viewed normally. By doing this, the π terms cancel and give $r_f = L/F$. These same publications often use I instead of L for radiance and write it as $r_f = I/F$.

5.2 Disk-Resolved Albedos and Measures of Reflectance

instances, $r_f < 1.0$, but this is not required. Some surfaces may reflect more light in a given direction than a Lambertian surface; in those directions, the radiance factor can exceed 1.0, but over all directions, the sum of energy must be conserved. Since we are dividing a radiance by a radiance, the radiance factor is dimensionless.

For a Lambertian surface, the radiance is only a function of cos i, giving

$$r_f = k \cos i = k\mu_0 \qquad 5.11$$

For a perfect Lambertian surface, $r_f = \mu_0$.

5.2.4 Radiance Coefficient, r_c

Also sometimes referred to as the *Lambert albedo* or *reflectance factor*, this common reflectance measurement is defined as the radiance of a sample at a given illumination and viewing geometry divided by that of a Lambertian surface illuminated and viewed *at the same* geometry.

$$r_c = \frac{L(i, e, \alpha)}{L_{Lam}(i, e, \alpha)} \qquad 5.12$$

For a Lambertian surface, the numerator and denominator have the same functional behavior, giving

$$r_c = k \qquad 5.13$$

For a perfect Lambertian surface, $k = 1$ and so $r_c = 1$.

The radiance coefficient gives one an intuitive sense of the reflectance properties of any given surface because it compares it to a Lambertian ideal under identical conditions. Unfortunately, it is trickier to measure in the laboratory because no real surfaces behave as ideal Lambertian reflectors at all geometries, so some type of correction is usually required. It can be derived relatively easily, and probably more accurately, by measuring the radiance factor of the sample in the lab and then converting to radiance coefficient using

$$r_c = \frac{r_f}{\mu_0} \qquad 5.14$$

Dividing by μ_0 (= cos i) thus mimics measuring relative to an ideal Lambertian standard. The radiance coefficient is also dimensionless. The radiance coefficient is related to the measures previously described by

$$r_c = \pi r_{brdf} = \frac{\pi r}{\mu_0} \qquad 5.15$$

5.2.5 Normal Albedo or Normal Reflectance, r_n

There are conflicting definitions for this term. Lester et al. (1979) define it to be the radiance of a surface illuminated and viewed perpendicularly (or normally), compared to the radiance of a perfect Lambertian surface also illuminated and viewed perpendicularly. This can be written as

$$r_n = \frac{L(0,0,0)}{L_{Lam}(0,0,0)} = r_f(0,0,0) = r_c(0,0,0) = \pi r(0,0,0) \qquad 5.16$$

For a Lambertian surface of albedo k, $r_n = k$.

Hapke (1981, 1993) defines it slightly differently as the radiance of a surface illuminated and viewed at 0° *phase angle*, compared to the radiance of a Lambertian surface illuminated and viewed perpendicularly. This subtle distinction means that, according to the Hapke definition, the surface in question may be viewed at non-zero incidence or emission angles so long as they are the same and result in $\alpha = 0°$.

$$r_n = \frac{L(e,e,0)}{L_{Lam}(0,0,0)} \qquad 5.17$$

Here we have indicated that the incidence and emission angle are non-zero and identical.

Defined in this (Hapke) way, the normal albedo is useful for comparing the intrinsic reflectance of different areas on the Moon, Mercury, asteroids, or any other atmosphereless solid body that is relatively dark. When viewed near $\alpha = 0°$ the incidence and emission angle increase as you look away from the sub-Earth point and toward the limb, but the phase angle is unchanged. Hapke (1993) shows that, at least for dark surfaces, differences in apparent brightness at this geometry are largely intrinsic to the material making up the planet. Both definitions given above are dimensionless.

5.2.6 Directional-Hemispherical Albedo or Reflectance, r_h

When an incident beam of light hits a non-specular surface, the energy is scattered or reflected into all directions. The directional-hemispherical reflectance is the ratio of the total reflected power, in all directions, to that incident.

The irradiance (power) incident on a unit element of surface is given (Eq. 2.14)

$$E = F \cos i = F\mu_0 \qquad 5.18$$

and reflects into a small area of solid angle $d\Omega$ (Figure 5.3) given by

5.2 Disk-Resolved Albedos and Measures of Reflectance

$$dA = R^2 d\Omega = R^2 (\sin e \, de \, d\phi) \qquad 5.19$$

Integrating the reflected radiances over the entire upper hemisphere (taking into account the projected area effect with cos e) gives the total power reflected from that unit area (sometimes called the exitance or emittance, M):

$$M = \iint L(e,\phi) \cos e \, d\Omega = \int_0^{2\pi} d\phi \int_0^{\pi/2} L(e,\alpha) \cos e \sin e \, de$$

$$M = 2\pi \int_0^{\pi/2} L(e,\alpha) \cos e \sin e \, de \qquad 5.20$$

The exitance has units of power per unit area, just as the irradiance. If we divide M by the incident irradiance, we can write the directional-hemispherical reflectance as

$$r_h = \frac{M}{\mu_0 F} = \frac{2\pi}{\mu_0} \int_0^{\pi/2} r(i,e,\alpha) \cos e \sin e \, de \qquad 5.21$$

or

$$r_h = 2\pi \int_0^{\pi/2} r_{brdf}(i,e,\alpha) \cos e \sin e \, de$$

or

$$r_h = \frac{2}{\mu_0} \int_0^{\pi/2} r_f(i,e,\alpha) \cos e \sin e \, de$$

or

$$r_h = 2 \int_0^{\pi/2} r_c(i,e,\alpha) \cos e \sin e \, de$$

If we assume the surface is Lambertian, by definition, $r_c = k$ (a constant) and the latter equation can be integrated to give

$$r_{h,L} = k \qquad 5.22$$

If the surface is a perfect Lambertian, then $r_{h,L} = 1$.

5.2.7 Photometric Function, $f(i, e, \alpha)$

The photometric function arose from a need to examine the change in brightness across any spherical surface, but historically across the surface of the Moon. Minnaert (1961) defines the photometric function to be the ratio of the radiance factor observed from a point on a sphere to that observed from the same surface when illuminated and observed perpendicularly (or normally) to the surface. What is different here is that the reference standard is the *surface itself* and not a perfect Lambertian surface. This has the effect of removing any intrinsic albedo of the surface, leaving only the angular behavior.

We can write the photometric function as

$$f(i,e,\alpha) = \frac{L(i,e,\alpha)}{L_{Lam}(0,0,0)} \div \frac{L(0,0,0)}{L_{Lam}(0,0,0)} = \frac{L(i,e,\alpha)}{L(0,0,0)} \qquad 5.23$$

or alternatively

$$r_f = r_n f(i,e,\alpha)$$

where $f(0, 0, 0) = 1.0$ and the constant r_n is the normal albedo. For a Lambertian surface with albedo k, the photometric function is

$$f(i,e,\alpha) = \frac{k \cos i}{k} = \mu_0 \qquad 5.24$$

It is the disk-resolved equivalent of the integral phase function, to be discussed later.

Several authors (e.g. Kreslavsky et al., 2000; Shkuratov et al., 2011) will write the photometric function as a product of two separate functions:

$$f(i,e,\alpha) = f(\Lambda, \Theta, \alpha) = f_o(\alpha) D(\Lambda, \Theta, \alpha) \qquad 5.25$$

Here, the photometric function has been rewritten in terms of luminance coordinates (as for the Moon where this type of analysis is most common; see Chapter 2). The first term, f_o, is often referred to as the *brightness phase function* and describes the change in brightness with phase independent of coordinates on the Moon. The term D is called the **disk function** and describes the brightness variations over the disk itself (Shkuratov et al., 2011). We will see an example of this in Section 5.4.

5.3 Reciprocity

A useful relationship for the bidirectional reflectance of a surface is known as the **Principle of Reciprocity**. Initially formulated by Helmholtz and described by Minnaert (1941), it is a restatement of the Law of Conservation of Energy.

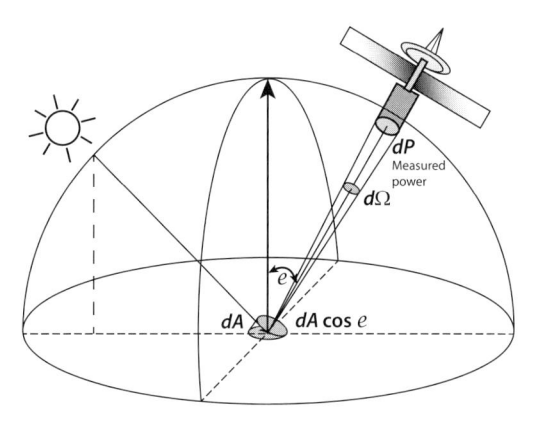

Figure 5.4. Scattering geometry.
Credit: Michael K. Shepard.

5.3.1 Derivation

Consider again the amount of light with incident flux on a small area of a surface, dA, at incidence angle i, emission angle e, and phase angle α (Figure 5.4). If we consider only the measured power per unit solid angle reflected from this area (the radiant intensity, I), conservation of energy requires this to be the same if we reverse the directions of incidence and emission.

$$I(i, e, \alpha) = I(e, i, \alpha) \qquad 5.26$$

If we divide both sides by their respective projected areas, we get the radiance for each reflectance geometry and note that they are not equal:

$$L(i, e, \alpha) = \frac{I(i, e, \alpha)}{dA \cos e} \neq L(e, i, \alpha) = \frac{I(e, i, \alpha)}{dA \cos i} \qquad 5.27$$

But if we now divide each reflected radiance by their respective irradiances, we get their bidirectional reflectance function, r_{brdf} and see that *these are equal*:

$$r_{brdf}(i, e, \alpha) = \frac{I(i, e, \alpha)}{F \cos i \, dA \cos e} = r_{brdf}(e, i, \alpha) = \frac{I(e, i, \alpha)}{F \cos e \, dA \cos i} \qquad 5.28$$

We can restate the principle of reciprocity for the other reflectance functions as well:

$$\begin{aligned} r(i, e, \alpha) \cos e &= r(e, i, \alpha) \cos i \\ r_f(i, e, \alpha) \cos e &= r_f(e, i, \alpha) \cos i \\ r_c(i, e, \alpha) &= r_c(e, i, \alpha) \end{aligned} \qquad 5.29$$

For a scattering model to be physically realistic, this principle must hold. Some empirical scattering relationships do not obey reciprocity but may still be useful when one is interested in deriving an empirical fit over a restricted set of illumination/viewing conditions.

5.3.2 Applications

The Principle of Reciprocity is an extremely powerful tool in laboratory and planetary photometry – laboratory data can be checked for consistency at different illumination and viewing geometries, and planetary surfaces can be directly compared at complementary geometries.

Minnaert (1941, 1961) illustrates the reciprocity symmetries that allow for comparisons between different parts of a planetary disk, as shown in Figure 5.5. The disk represents the Earth-based view of a spherical planet, and the sub-Earth point is always the center of the disk. The vertical center line through the disk is the **viewing** or **central meridian.** The emission vector extends from the center of the sphere, through the point on the surface at the center of the disk, to the observer. At the center point, the emission angle is 0°, and it increases radially away from the center until it is 90° at the limb. The points labeled **S** represent the sub-solar point and indicate the point on the surface where the incident solar light is perpendicular or normal to the surface. The angular distance between the sub-Earth and sub-solar points is the phase angle, and the great circle line between these two points is the **luminance** or **radiance equator.** If there are subscripts, it indicates sub-solar points mirrored about the central meridian. The points labeled **P** are areas on the surface where we are interested in the reflected light properties.

For the purposes of this discussion, we assume the spherical surface is of uniform composition and there is little or no macroscopic roughness to cast shadows. And for ease of discussion, we will use the term *reflectance* to specifically mean the r_{brdf}.

In Figure 5.5a, we show the mirror meridian, a meridian that bisects the radiance equator between sub-Earth and sub-solar points. Equidistant from the mirror meridian at each latitude, there will be two points, **P**; the only difference in their photometric position is that the incidence and emission angles have been swapped. In this case, reciprocity tells us that the reflectance of these two points must be the same. If they are not, then one of our assumptions is incorrect, the most likely being the assumptions of identical composition or identical surface texture.

In Figure 5.5b, we see that for any given phase angle, two points at mirror latitudes and the same longitude will have identical reflectance. Both points have identical incidence, emission, and phase angles. In

5.3 Reciprocity

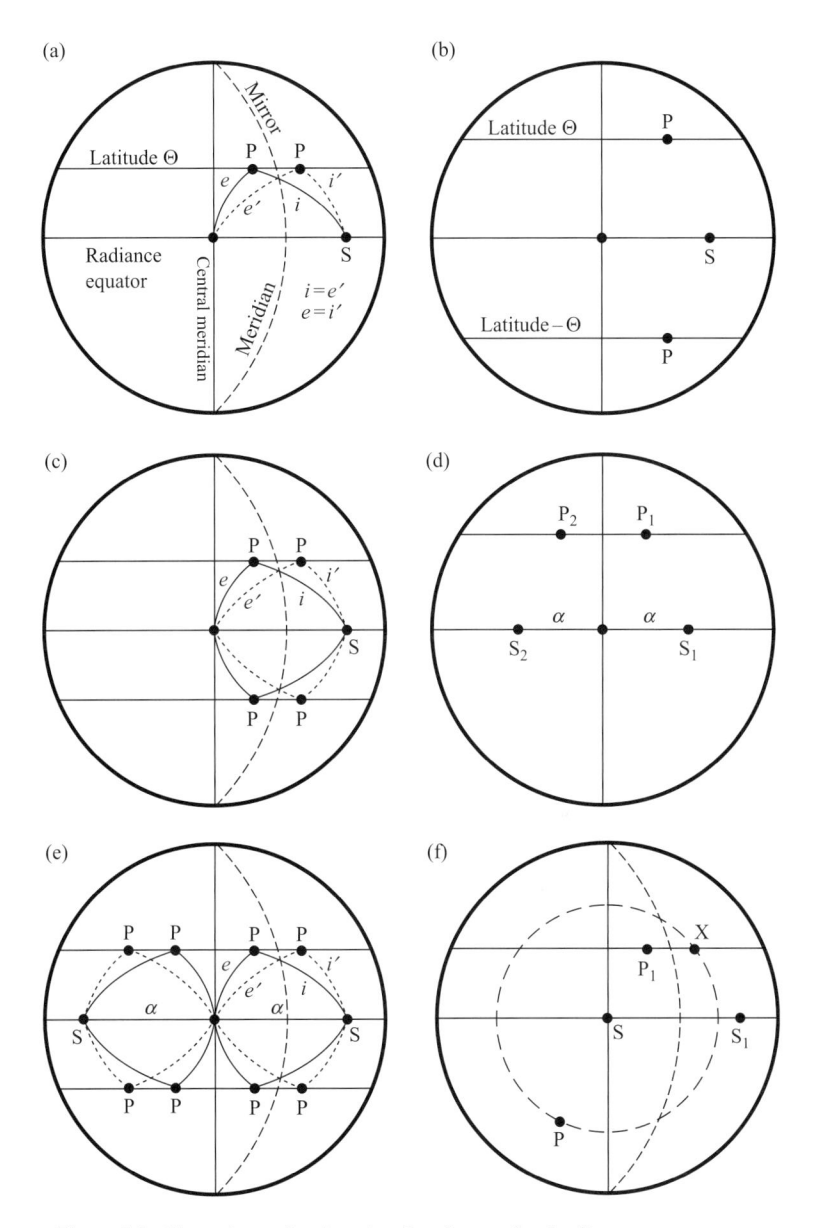

Figure 5.5. Illustrations of reciprocity. See the text for details.
Credit: After Minnaert (1941).

Figure 5.5c, we combine the results of (a) and (b) to show that, for any given phase angle, four points may be directly compared.

In Figure 5.5d, we consider a point $\mathbf{P_1}$ at latitude Θ viewed with the Sun at $\mathbf{S_1}$ and phase angle α. When the Sun has moved so that it is now at $\mathbf{S_2}$, also at phase angle α, the mirror of $\mathbf{P_1}$ about the central meridian point, $\mathbf{P_2}$, also at latitude Θ, will have the same reflectance. Both points have identical incidence, emission, and phase angles. A special case occurs when the two points coincide on the central meridian and gives a useful way of checking the calibration of images taken on different days. In Figure 5.5e, we combine the rules from (a), (b), and (d) to show that eight points can be compared for any two mirror phase angles.

Finally, in Figure 5.5f we consider the special case of sphere at opposition (full moon); the sub-solar point is coincident with the sub-Earth point at the center of the disk. For any circle concentric with the disk perimeter or limb, all points on the circle with $\mathbf{P}$ have identical incidence and emission angles and all have the same reflectance, assuming the initial assumptions for identical composition and texture hold. In this case, the point labeled $\mathbf{X}$ has the same reflectance as $\mathbf{P}$.

Using the methods outlined previously, we can also compare the reflectance of any two points, $\mathbf{P}$ and $\mathbf{P_1}$, also shown in Figure 5.5f. As noted, point $\mathbf{X}$ on the dashed circle has the same reflectance as $\mathbf{P}$ when $\alpha = 0°$. It was chosen because it is at the same latitude as $\mathbf{P_1}$. By using the method of reciprocity illustrated in Figure 5.5a, the reflectance of $\mathbf{X}$ will be the same as $\mathbf{P_1}$ when the Sun is at $\mathbf{S_1}$. And one last time, we reiterate that these principles only hold when the material at two different points is the same. When they don't hold, it tells us something about the albedo or textural differences of the two points.

5.4 Disk-Resolved Scattering Models and Behavior

In this section, we examine a variety of disk-resolved scattering models. A summary table of the more important equations is given in Table 5.1.

5.4.1 Limb Darkening and Brightening

An early goal of planetary astronomers was to use the apparent change in scattering behavior across the disk of a planet to derive its scattering model. For any given view, the phase angle remains a constant (because neither the position of the Sun nor observer change), but the local

5.4 Disk-Resolved Scattering Models and Behavior

incidence and emission angle change as your eye scans across the disk to different luminance latitudes and longitudes. Note that this assumes the observer is standing far off. Unfortunately, we now know that observations at one phase angle are not sufficient to provide a unique scattering model (Harris, 1961). However, these observations can still provide some useful information.

When viewed near zero-phase angle (i.e. the entire disk is illuminated and viewed) there are three possible conditions: the disk may be uniformly bright, **limb-darkened**, or **limb-brightened**. These behaviors are still useful as descriptors.

5.4.2 Lambertian

We have derived all the major reflectance quantities for Lambertian surfaces in a previous section and will not repeat them here. We will, however, note that a Lambertian scattering model is a reasonable one for many bright surfaces, like the icy satellites of Jupiter or Saturn, is extremely simple to implement, and is physically plausible because it obeys reciprocity:

$$r_f(i,e,\alpha)\cos e = k\cos i \cos e$$
$$r_f(e,i,\alpha)\cos i = k\cos e \cos i \qquad 5.30$$

An important property of the Lambertian scattering model is that spheres covered by material obeying this model are limb-darkened.

Using the relationships between surface and luminance coordinates, the Lambertian model can be written

$$r = \frac{\cos i}{\pi} = \frac{1}{\pi}\cos(\Lambda + \alpha)\cos\Theta \qquad 5.31$$

At opposition, as either latitude or longitude plus phase angle approach 90°, the reflectance drops to 0. For any phase, the disk is limb darkened at extreme latitudes. Additionally, the terminator occurs at $\Lambda = 90° - \alpha$; inserting this into the preceding equation shows that a Lambert sphere is also limb-darkened as one approaches it. Many of the bright, icy satellites are Lambertian and show limb darkening. Planets with thick atmospheres (e.g. Uranus and Venus) also show limb darkening and masquerade as Lambertian surfaces (Figure 5.6).

One major shortcoming of the Lambert scattering model is that it is not accurate for the darker surfaces in the solar system, like those of Mercury and the Moon. For that, we turn to the Lommel–Seeliger scattering model.

Figure 5.6. Examples of planetary limb darkening. Top (L-R) Uranus ($\alpha \sim 0°$) and Enceladus ($\alpha \sim 35°$) are examples of objects covered with thick atmospheres or have bright surfaces that approximate Lambertian behavior. Bottom (L-R) Moon ($\alpha \sim 0°$) and Mercury ($\alpha \sim 35°$) show surfaces that approximate Lommel–Seeliger scattering. Compare with simulations in Figure 5.9.

Credit: Moon image credit: NASA/Goddard/LRO: www.nasa.gov/feature/goddard/rare-full-moon-on-christmas-day; Mercury image credit: NASA/JHUAPL/Carnegie Institute of Washington. http://photojournal.jpl.nasa.gov/catalog/PIA11245; Uranus image credit NASA/ESA, Erich Karkoschka, University of Arizona: www.spacetelescope.org/images/opo0405d/; Enceladus image credit: NASA/JPL-Caltech/Space Science Institute. http://photojournal.jpl.nasa.gov/catalog/PIA17202

5.4.3 Lommel–Seeliger

Also referred to as the **Lunar Law**, this scattering model is one of the most commonly used in planetary applications. It is named for two men who played significant roles in the history of planetary photometry and appear to have independently derived the equation. **Eugen von Lommel** (1837–1899) was a German physicist who was instrumental in the development of the theory of radiative transfer (Mischenko, 2013), and Hugo von Seeliger (Chapter 4) is famous for his explanation of the rapid brightening of Saturn's rings at opposition.

5.4 Disk-Resolved Scattering Models and Behavior

As shown in the previous chapter, the Lommel–Seeliger scattering model is the simplest expression that can be derived from the equation of radiative transfer. It is reasonably accurate only for dark surfaces. It may be expressed in the following forms:

$$r = \frac{w}{4\pi} \frac{\mu_0}{\mu + \mu_0}$$
$$r_{brdf} = \frac{w}{4\pi} \frac{1}{\mu + \mu_0}$$
$$r_f = \frac{w}{4} \frac{\mu_0}{\mu + \mu_0}$$
$$r_c = \frac{w}{4} \frac{1}{\mu + \mu_0}$$
5.32

Here, w is the particle single-scattering albedo, the ratio of the amount of light that scatters from a particle to the total incident light, and $0 \leq w \leq 1$. In luminance coordinates, we can write (using Eqs. 2.5 and 2.6)

$$r_{brdf} = \frac{w}{4\pi} \frac{1}{(\cos(\alpha + \Lambda) + \cos\Lambda)\cos\Theta}$$
5.33

or

$$r = \frac{w}{4\pi} \frac{\cos(\alpha + \Lambda)}{\cos(\alpha + \Lambda) + \cos\Lambda}$$

Note that the latter version has no dependence on luminance latitude, Θ.

When observed at opposition ($\alpha = 0°$), we see that the bidirectional reflectance is a constant across the disk. The only variations in brightness to be observed would be due to changes in the makeup of the surface. For phases other than opposition, it is apparent that as we approach the terminator ($\Lambda \geq 90° - \alpha$), the reflectance darkens dramatically. And at the limb ($\Lambda = -90°$ in our geometry), the surface is brightest (limb-brightening) because the $\cos \Lambda$ term in the denominator goes to 0.

A quick check shows that the L-S law obeys reciprocity:

$$r_f(i, e, \alpha) \cos e = \frac{w}{4} \frac{\mu_0 \mu}{\mu_0 + \mu}$$
$$r_f(e, i, \alpha) \cos i = \frac{w}{4} \frac{\mu \mu_0}{\mu + \mu_0}$$
5.34

The normal albedo of an L-S surface is

$$r_n = \pi r(0, 0, 0) = \frac{w}{4} \frac{1}{1 + 1} = \frac{w}{8}$$
5.35

Since $w \leq 1$, the normal reflectance of a L-S surface must be $r_n \leq 0.125$.

The directional-hemispherical albedo of an L-S surface is

$$r_h = \frac{2\pi}{\cos i} \int_0^{\pi/2} \frac{w}{4\pi} \frac{\cos i}{\cos i + \cos e} \cos e \sin e \, de = \frac{w}{2} \int_0^{\pi/2} \frac{\cos e \sin e}{\cos i + \cos e} de \quad 5.36$$

The solution to this integral is

$$r_h = \frac{w}{2} \left[1 - \mu_0 \ln\left(1 + \frac{1}{\mu_0}\right) \right] \quad 5.37$$

The photometric function of a Lommel–Seeliger surface can be found from the relation

$$f(i, e, \alpha) = \frac{r_f}{r_n} = \frac{w}{4} \frac{\mu_0}{\mu + \mu_0} \frac{8}{w} = \frac{2\mu_0}{\mu + \mu_0} \quad 5.38$$

A quick check of this will show that, at opposition, $f(i, e, 0) = 1$, i.e. it is flat across the disk.

5.4.3.1 Generalized Lommel–Seeliger

The Lommel–Seeliger model can be easily modified to account for non-isotropic scattering. In the derivation in Chapter 4, instead of assuming the light is scattered equally in all directions, we may specify a phase function $P(\alpha)$ that describes the directional dependence of scattering:

$$L(\tau, \mu_0, \mu, \alpha) = \frac{wE}{\mu 4\pi} P(\alpha) e^{-\frac{\tau}{\mu_0}} \quad 5.39$$

This leads to the following bidirectional reflectance function

$$r = \frac{w}{4\pi} \frac{\mu_0}{\mu + \mu_0} P(\alpha) \quad 5.40$$

In luminance coordinates, the generalized Lommel–Seeliger law can be rewritten

$$r = \frac{w}{4\pi} \frac{\cos(\alpha + \Lambda)}{\cos(\alpha + \Lambda) + \cos \Lambda} P(\alpha) \quad 5.41$$

5.4.4 Lunar–Lambert

To approximate the scattering behavior of real surfaces, the Lommel–Seeliger (Lunar) and Lambertian models are often combined into a hybrid model called the **Lunar–Lambert**. The Lambertian works better for bright surfaces,

5.4 Disk-Resolved Scattering Models and Behavior

while the Lommel–Seeliger works best (albeit only fairly) for dark surfaces. Blending the two models provides more flexibility in fitting intermediate cases. The simplest blending model is of the form

$$r \text{ or } r_f = A\frac{\mu_0}{\mu + \mu_0} + B\mu_0 \qquad 5.42$$

where A and B can be constants or more complex functions of phase angle if desired. Note that both the "albedo" and weighting functions are intrinsic to A and B. In some applications of this model, A and B are constrained so that $A + B = 1$; in others, they are not. Similar forms can be written for the bidirectional reflectance distribution function or radiance coefficient:

$$r_{brdf} \text{ or } r_c = A\frac{1}{\mu + \mu_0} + B \qquad 5.43$$

A quick check shows that the Lunar–Lambert model, like its individual components, obeys reciprocity.

In luminance coordinates, we can write this law as

$$r = A\frac{\cos(\alpha + \Lambda)}{\cos(\alpha + \Lambda) + \cos\Lambda} + B\cos(\alpha + \Lambda)\cos\Theta \qquad 5.44$$

5.4.5 Minnaert

This is a generalized Lambertian model suggested by Minnaert (1941) of the form

$$r_f(i, e, \alpha) = k\,\mu_0^n \mu^{n-1} \qquad 5.45$$

or in luminance coordinates

$$r_f = k\,[\cos(\alpha + \Lambda)\cos\Theta]^n [\cos\Lambda\cos\Theta]^{n-1} \qquad 5.46$$

Note that for $n = 1$, this reduces to a Lambertian model.

Although it obeys reciprocity,

$$\begin{aligned} r_f(i, e, \alpha)\mu &= k\,\mu_0^n \mu^{n-1} \mu = k\,\mu_0^n \mu^n \\ r_f(e, i, \alpha)\mu_0 &= k\,\mu^n \mu_0^{n-1} \mu_0 = k\,\mu^n \mu_0^n \end{aligned} \qquad 5.47$$

the Minnaert function has had mixed success. The biggest issue seems to be that, while it may fit a subset of data well, it behaves badly for data sets with a large range of incidence, emission, and phase angles, i.e. it can't reproduce full scattering curves accurately. It also has trouble at the limbs where $e = 90°$. Hapke (2012) points out that for $n < 1$ the reflectance becomes infinite at the limb, while for $n > 1$ it goes to zero. Nevertheless, it has been used with success in many studies where the range of observed phase angles is restricted.

5.4.6 Akimov Disk Function

Historically, a great deal of effort has been expended in attempting to find a simple photometric model that fit the observed disk function behavior of the Moon. The Lunar–Lambert and Minnaert generally have had the most application, especially in the West, but there are two models proposed by Akimov (1979, 1988) that are also used (see discussion by Shkuratov et al., 2011). The first is strictly empirical:

$$D(\Lambda, \Theta, \alpha) = \frac{\left[\cos^{q+1}\left(\Lambda - \frac{\alpha}{2}\right) - \sin^{q+1}\left(\frac{\alpha}{2}\right)\right]}{\cos \Lambda \left[1 - \sin^{q+1}\left(\frac{\alpha}{2}\right)\right]} \cos^{q}\Theta \qquad 5.48$$

where $q = 0.3\alpha$ and α is in radians. The second is theoretically derived

$$D(\Lambda, \Theta, \alpha) = \frac{\cos\left[\frac{\pi}{\pi-\alpha}\left(\Lambda - \frac{\alpha}{2}\right)\right]}{\cos \Lambda} (\cos \Theta)^{\frac{\alpha}{\pi-\alpha}} \qquad 5.49$$

These functions have proven to be very useful for normalizing images of the lunar disk and removing large-scale albedo variations. Normalized images acquired at different phase angles are then compared, and the resulting ratio images are very sensitive to photometric anomalies on the lunar surface (Kreslavsky and Shkuratov, 2003; Kaydash et al., 2010; Shkuratov et al., 2010, 2011; Clegg et al., 2014).

5.5 Disk-Integrated Albedos and Measures of Reflectance

In this section, we examine the types of albedo or reflectance for planets as a whole. In the previous section, we relied heavily upon the definition of radiance, which involved the power per unit area and solid angle. In this section, we are not measuring power per area, but only per solid angle, so a more convenient parameter to work with is the radiant intensity, I (Chapter 2).

5.5.1 Disk Integration

Before beginning, we derive the fundamental relationship between the radiant intensity, I, reflected from a sphere and the geometry of the sphere from Chapter 2. For some small area on the sphere, we can write the area, da, as a function of position and the planetary radius, R:

5.5 Disk-Integrated Albedos and Measures of Reflectance

$$da = R^2 \cos \Theta \, d\Theta \, d\Lambda \qquad 5.50$$

If an incident beam of flux density, F, strikes the sphere, each area on the surface will reflect some of that power. We can calculate the radiant intensity from that small area with

$$dI = (F \cos i) \, (r_{brdf}) \, (\cos e R^2 \cos \Theta \, d\Theta \, d\Lambda) \qquad 5.51$$

and we have grouped the terms to make it easier to see their function: the first is the irradiance, the second is the reflected radiance, and the final term is the area with the projection effect taken into account so that we are left with radiant intensity (power per unit solid angle).

To find the radiant intensity over the entire sphere, we regroup terms, substitute Equations 2.5 and 2.6 in for $\cos i$ and $\cos e$, and then integrate over all visible latitudes and longitudes

$$I = R^2 F \iint r_{brdf} \cos(\alpha + \Lambda) \cos \Theta \cos \Lambda \cos \Theta \cos \Theta \, d\Theta \, d\Lambda$$

$$I = R^2 F \int_{\Theta=-\frac{\pi}{2}}^{\frac{\pi}{2}} \int_{\Lambda=-\frac{\pi}{2}}^{\frac{\pi}{2}-\alpha} r_{brdf} \cos^3 \Theta \cos(\alpha + \Lambda) \cos \Lambda \, d\Theta \, d\Lambda \qquad 5.52$$

Note the limits on the integrations; all latitudes, Θ, are visible, but the longitudes cease at the terminator, given by $\Lambda = \pi/2 - \alpha$ (Figure 5.7).

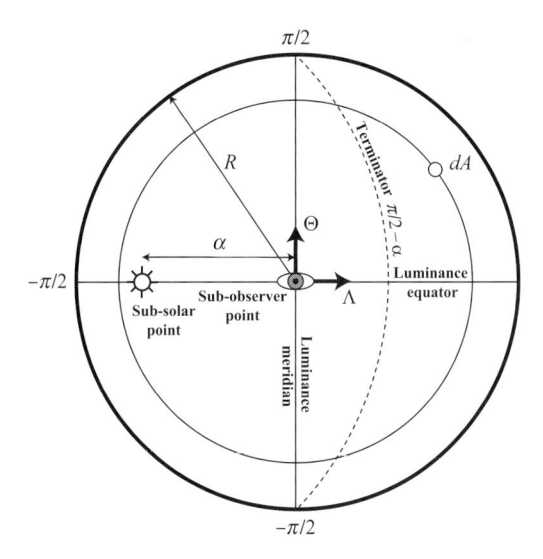

Figure 5.7. Geometry for spherical integration using luminance coordinates. Credit: Michael K. Shepard.

For the special case of opposition ($\alpha = 0$), this function simplifies to

$$I(0) = R^2 F \int_{\Theta=-\frac{\pi}{2}}^{\frac{\pi}{2}} \int_{\Lambda=-\frac{\pi}{2}}^{\frac{\pi}{2}} r_{brdf} \cos^3\Theta \cos^2\Lambda \, d\Theta \, d\Lambda \qquad 5.53$$

Hapke (1993) takes advantage of the symmetry of the full disk and writes this only in terms of emission angle

$$\begin{aligned} I(0) &= F \int_0^{2\pi} d\phi \int_0^{\pi/2} r(e,e,0) \cos e \, R \sin e \, R \, de \\ &= 2\pi R^2 F \int_0^{\pi/2} r(e,e,0) \cos e \sin e \, de \qquad 5.54 \end{aligned}$$

When looking at a disk at 0° phase angle, all points have identical incidence and emission angles (Figure 5.8). This allows us to write the bidirectional reflectance as only a function of emission angle. This version is simpler to integrate. Note however, Hapke's version uses the bidirectional reflectance, while the previous uses the bidirectional reflectance distribution function.

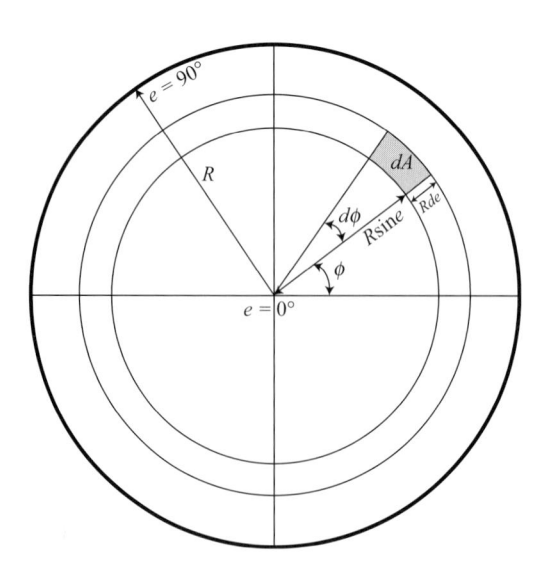

Figure 5.8. Geometry for spherical integration using the method of Hapke.
Credit: Michael K. Shepard.

5.5 Disk-Integrated Albedos and Measures of Reflectance

5.5.2 Geometric Albedo, p

Also referred to as the **physical albedo** (thus the common symbol, p) this is the amount of light – in all wavelengths – reflected by a planetary body divided by the amount that would have been reflected by a perfect Lambertian disk of the same *cross-sectional area*, located at the same distance. Both the real object and the imaginary Lambertian object must be illuminated and viewed at 0° phase angle, i.e. at opposition.

In most planetary applications, the intent is to measure the **visual geometric albedo**, p_v. This differs from the general definition in that it is the amount of light reflected in the *visual* wavelengths, usually defined as the passband of a Johnson-Cousins visual filter which filters out all light except that between roughly 500–600 nm. It is common practice to use the term **geometric albedo** in this more restricted sense of wavelengths.

Note that we are NOT comparing the reflectance of our planet to an equally sized and shaped planet composed of Lambertian material, but instead to a two-dimensional cross-section or cut-out disk made of Lambertian material. We can write

$$p = \frac{I(0)}{I_{Lam}} \qquad 5.55$$

where $I(0)$ is the radiant intensity from the planet at $\alpha = 0°$ and I_{Lam} is the radiant intensity of the Lambertian disk, also at $\alpha = 0°$. For a Lambertian disk of radius, R

$$I_{Lam} = L_{Lam}\pi R^2 = \frac{F(\pi R^2)}{\pi} = FR^2 \qquad 5.56$$

(units of W sr^{-1}) so that

$$p = \frac{I(0)}{FR^2}$$

5.5.3 Integral Phase Function, $\Phi(\alpha)$

This is the function that describes the directional dependence of light scattered in all directions by a planet. It is the brightness (radiant intensity) of the entire disk seen at some phase angle, α, divided by the brightness of the disk at $\alpha = 0°$.

$$\Phi(\alpha) = \frac{I(\alpha)}{I(0)} \qquad 5.57$$

Because the radiant intensity is highest at $\alpha = 0°$, the integral phase function is normalized, so that $\Phi(0°) = 1.0$ and $\Phi(\alpha > 0°) < 1.0$. The form of this function suggests that it is a disk-integrated analog of the disk-resolved photometric or disk function, defined earlier.

5.5.4 Spherical or Bond Albedo and Phase Integral

The spherical or Bond albedo (named for **George Bond**, 1825–1865) is the fraction of total incident power – all wavelengths – that is scattered into space and not absorbed by the planet.

$$A = \frac{P_{scattered}}{P_{incident}} \qquad 5.58$$

The total power incident on the planet is

$$P_{incident} = \pi R^2 F \qquad 5.59$$

where R is the radius of the planet and F is the incident collimated irradiance from the Sun. Assuming azimuthal independence, the total amount scattered by the planet can be found from

$$P_{scattered} = \int_0^{4\pi} I(\alpha) d\Omega = \int_0^{2\pi} d\phi \int_0^{\pi} I(\alpha) \sin\alpha \, d\alpha = 2\pi \int_0^{\pi} I(\alpha) \sin\alpha \, d\alpha \qquad 5.60$$

Dividing this by the total incident power gives us the Bond albedo.

$$A = \frac{2\pi \int_0^{\pi} I(\alpha) \sin\alpha \, d\alpha}{\pi R^2 F} = \frac{2 \int_0^{\pi} I(\alpha) \sin\alpha \, d\alpha}{R^2 F} \qquad 5.61$$

Multiply top and bottom by $I(0)$ and rearrange to get

$$A = \frac{I(0)}{R^2 F} 2 \int_0^{\pi} \frac{I(\alpha)}{I(0)} \sin\alpha \, d\alpha = \frac{I(0)}{R^2 F} 2 \int_0^{\pi} \Phi(\alpha) \sin\alpha \, d\alpha \qquad 5.62$$

The first term is the geometric albedo (Eq. 5.56). The second term is a function only of the scattering and shape properties of the planetary surface and defines the **phase integral**, q. In essence, this is the weighted average of scattering over all phase angles.

$$q = 2 \int_0^{\pi} \Phi(\alpha) \sin\alpha \, d\alpha \qquad 5.63$$

$$A = pq$$

The Bond albedo tells us how much incident light scatters from the surface, and by inference, how much is absorbed. This is useful for calculating heat balances – how hot will a planet get given its illumination? It is also the planetary equivalent of the particle single-scattering albedo, w.

5.5.5 Russell's Law

Russell's "law" or rule is an empirical observation, first made by the astronomer Henry Norris Russell (1916b), that there is a relationship between the phase integral and integral phase function:

$$q \approx 2.2\,\Phi(50°) \qquad 5.64$$

The rule is found to hold to within 5% for the terrestrial planets and asteroids. Veverka (1971a) shows that this relationship is a simple consequence of the scattering behavior of surfaces with backscattering behavior, such as the asteroids, Moon, Mercury, and Mars. For Venus and its thick, forward scattering atmosphere, the rule still holds, but not quite as well. The rule is useful because it is rare to measure the entire phase integral of a planet, a necessity if we wish to integrate over all phase angles to obtain the integral phase function. Here we see that it can be estimated to a good approximation if we can measure the phase function at one geometry.

5.6 Disk-Integrated Scattering Models and Behavior

Until recently, the planets were essentially points of light. Telescopes allowed the disks to be seen, but much was done by looking only at the integrated brightness of a planet as its phase angle changed. In this section, we will examine the disk-integrated behavior of the Lambert and Lommel–Seeliger models, as these were the most commonly used. We will also examine a physically unrealizable but useful scattering law called the Area Law (Figure 5.9). Summaries of the most important equations are given in Table 5.2.

5.6.1 Lambertian

Spheres covered by brighter surfaces displaying Lambertian behavior, or shrouded by thick atmospheres, show significant limb darkening. The gas giants and some icy moons are examples of objects with Lambert-like limb-darkening behavior (Figures 5.8 and 5.9).

Figure 5.9. Illustration of whole disk scattering for the Lambert (bottom), Lommel–Seeliger (middle), and Area (top) scattering laws. Phase angles are 0°, 45°, 90°, and 135°, left to right. Each image has been stretched to show its full dynamic range.
Credit: Michael K. Shepard.

The intensity of light reflected from a Lambertian sphere can be found by substituting the bidirectional reflectance distribution function for a Lambertian surface into Equation 5.52.

$$I(\alpha) = R^2 F \int_{\Theta=-\frac{\pi}{2}}^{\frac{\pi}{2}} \int_{\Lambda=-\frac{\pi}{2}}^{\frac{\pi}{2}-\alpha} \frac{k}{\pi} \cos^3\Theta \cos(\alpha + \Lambda) \cos\Lambda \, d\Theta \, d\Lambda \qquad 5.65$$

To find the geometric albedo, we set $\alpha = 0$ and divide by the radiant intensity of a Lambertian disk of the same area (Eq. 5.56):

$$p = \int_{\Theta=-\frac{\pi}{2}}^{\frac{\pi}{2}} \int_{\Lambda=-\frac{\pi}{2}}^{\frac{\pi}{2}} \frac{k}{\pi} \cos^3\Theta \cos^2\Lambda \, d\Theta \, d\Lambda = \frac{2}{3}k = \frac{2}{3}r_{h,L} \qquad 5.66$$

where $r_{h,L}$ is the directional hemispherical reflectance of a Lambertian surface (see Eq. 5.22). For a perfect Lambertian surface, $p = 2/3$.

5.6 Disk-Integrated Scattering Models and Behavior

Alternatively, the Hapke version of the same integral is

$$p = 2\pi \int_0^{\pi/2} \frac{k}{\pi} \cos^2 e \sin e \, de = \frac{2}{3} k \qquad 5.67$$

To find the integral phase function, we must solve Equation 5.65 for $I(\alpha)$. The solution of this integral is straightforward, if tedious, and results in

$$I(\alpha) = \frac{2k}{3\pi} R^2 F [\sin \alpha + (\pi - \alpha) \cos \alpha] \qquad 5.68$$

Division by $I(0)$ gives the integral phase function

$$\Phi(\alpha) = \frac{I(\alpha)}{I(0)} = \frac{1}{\pi} [\sin \alpha + (\pi - \alpha) \cos \alpha] \qquad 5.69$$

and we may use this result to find the phase integral, q:

$$q = \frac{2}{\pi} \int_0^\pi [\sin \alpha + (\pi - \alpha) \cos \alpha] \sin \alpha \, d\alpha = \frac{3}{2} \qquad 5.70$$

Finally, since the Bond albedo $A = pq$ (Eq. 5.63), it is straightforward to use the previous results to get

$$A = \frac{2}{3} k \left(\frac{3}{2}\right) = k = r_{h,L} \qquad 5.71$$

Recall that the Bond albedo is the fraction of incident light scattered in all directions. If the sphere is a perfect Lambertian, it scatters all the incident light giving $A = 1$.

5.6.2 Lommel–Seeliger

Spheres covered by Lommel–Seeliger material have a flat or uniform brightness across most of their disk but show limb *brightening* at phase angles greater than 0° and darkening at the terminator. With the exception of the brightening at the limb, most of the darker, atmosphere-free objects, like the Moon, Mercury, and asteroids are good examples of Lommel–Seeliger-like behavior (Figures 5.6 and 5.9).

Merging Equations 5.32, 5.54, and 5.56 the geometric albedo of a Lommel–Seeliger sphere is given by[2]

---

[2] As pointed out by Fairbairn (2005), there is a persistent error in the literature for this quantity. Early writers, beginning with Russell (1916b) and continuing through Lester et al. (1979), made a mistake and derive the L-S value to be $p = w/4$.

$$p = 2\pi \frac{w}{4\pi} \int_0^{\frac{\pi}{2}} \frac{\cos e}{\cos e + \cos e} \cos e \sin e \, de = \frac{w}{8} \quad 5.72$$

To find the integral phase function, we first find $I(\alpha)$. Using Equations 5.33 and 5.52, we get

$$I(\alpha) = R^2 F \frac{w}{4\pi} \int_{\Theta=-\frac{\pi}{2}}^{\frac{\pi}{2}} \int_{\Lambda=-\frac{\pi}{2}}^{\frac{\pi}{2}-\alpha} \frac{\cos^2\Theta \cos(\alpha+\Lambda)\cos\Lambda}{\cos(\alpha+\Lambda)+\cos\Lambda} d\Theta \, d\Lambda \quad 5.73$$

$$I(\alpha) = R^2 F \frac{w}{4\pi} \int_{-\frac{\pi}{2}}^{\frac{\pi}{2}} \cos^2\Theta \, d\Theta \int_{-\frac{\pi}{2}}^{\frac{\pi}{2}-\alpha} \frac{\cos(\alpha+\Lambda)\cos\Lambda}{\cos(\alpha+\Lambda)+\cos\Lambda} d\Lambda$$

The solution to this expression is, again, straightforward but tedious and results in

$$I(\alpha) = R^2 F \frac{w}{8} \left[1 + \sin\frac{\alpha}{2} \tan\frac{\alpha}{2} \ln\left(\tan\frac{\alpha}{4}\right)\right] \quad 5.74$$

The integral phase function is then

$$\Phi(\alpha) = 1 + \sin\frac{\alpha}{2} \tan\frac{\alpha}{2} \ln\left(\tan\frac{\alpha}{4}\right) \quad 5.75$$

Despite individual terms blowing up at $\alpha = 0°$ (logarithm term) and $\alpha = 180°$ (tangent term), the function is well behaved over all phase angles.

Inserting this into the expression for the phase integral gives

$$q = 2 \int_0^\pi \left[1 + \sin\frac{\alpha}{2} \tan\frac{\alpha}{2} \ln\left(\tan\frac{\alpha}{4}\right)\right] \sin\alpha \, d\alpha = \frac{16}{3}(1 - \ln 2) = 1.637 \quad 5.76$$

The Bond albedo is then

$$A = pq = \frac{w}{8} \frac{16}{3}(1 - \ln 2) = 0.205 \, w \quad 5.77$$

5.6.3 Area Law

The so-called Area Law is a useful fiction described by Lester et al. (1979) that assumes that a surface obeys the scattering law

$$r_{brdf} = \frac{k}{\mu_0} \quad 5.78$$

5.6 Disk-Integrated Scattering Models and Behavior

or equivalently

$$r = k = \frac{r_f}{\pi} = \frac{\mu_0 r_c}{\pi}$$

The law is useful because spheres with surfaces that obey this scattering function have a flat brightness across their disk – there is no limb darkening or brightening. However, it does not obey reciprocity (recall, $r = k$):

$$r(i, e, \alpha) \cos e \neq r(e, i, \alpha) \cos i \qquad 5.79$$

Nevertheless, the model allows one to easily estimate the brightness of an object at any given phase angle because it is directly proportional to the visible area, and thus is called the Area Law.

We can find the geometric albedo of an object obeying this law with (remembering that $i = e$ in this geometry)

$$p = 2\pi \int_0^{\frac{\pi}{2}} k \cos e \sin e \, de = \pi k \qquad 5.80$$

The angular behavior of intensity is given by

$$I(\alpha) = R^2 F k \int_{\Theta=-\frac{\pi}{2}}^{\frac{\pi}{2}} \int_{\Lambda=-\frac{\pi}{2}}^{\frac{\pi}{2}-\alpha} \frac{\cos^3 \Theta \cos(\alpha+\Lambda)\cos\Lambda}{\cos(\alpha+\Lambda)\cos\Theta} d\Theta \, d\Lambda \qquad 5.81$$

where we have substituted for $\cos i$ in luminous coordinates. After canceling terms, we get

$$I(\alpha) = R^2 F k \int_{\Theta=-\frac{\pi}{2}}^{\frac{\pi}{2}} \int_{\Lambda=-\frac{\pi}{2}}^{\frac{\pi}{2}-\alpha} \cos^2 \Theta \cos\Lambda \, d\Theta \, d\Lambda = \frac{\pi}{2} R^2 F k (1 + \cos\alpha) \qquad 5.82$$

The integral phase function is found by dividing by $I(0)$ and gives

$$\Phi(\alpha) = \frac{1 + \cos\alpha}{2} \qquad 5.83$$

This is simply the area of the visible part of the sphere. Using this to solve for the phase integral gives

$$q = \int_0^{\pi} (1 + \cos\alpha) \sin\alpha \, d\alpha = 2 \qquad 5.84$$

From this and the geometric albedo, we find the Bond albedo $A = 2\pi k$.

5.7 More Complex Photometric Models

The models introduced so far have either been completely empirical or assumed that the surface is smooth, continuous, and made of a single homogeneous type of material. Unfortunately, these models fail to account for discontinuous particulate surfaces, scattering between particles, rough surfaces, or the widely observed opposition effect phenomenon.

As in other physics-oriented disciplines, modeling is often a contest between adding sufficient complexity to accurately describe observed behavior while keeping the expressions simple enough to use. In this section, we introduce two more recent and complex photometric models that address these complications. While there are several other scattering models in the literature, these two are the most frequently encountered.

5.7.1 Hapke

The Hapke model is ubiquitous in modern planetary photometry. Bruce Hapke first began to formulate a scattering model based on radiative transfer in 1963 to explain the scattering properties of the Moon. The fully realized model was published in Hapke (1981). The model has undergone numerous additions and revisions in the subsequent 35 years as new laboratory and spacecraft observations have become available. Because of all these modifications, it is more accurate to speak of Hapke models. In this book, we will focus on the version that is most commonly used and was developed in papers written from 1981–1986. In what follows, we use the nomenclature of the original papers and book (Hapke, 1993, 2012).

5.7.1.1 Multiple Scattering

The Hapke model differs from the more general Lommel–Seeliger model (Eq. 5.40) in only a few ways. First, multiple scattering is taken into account. With only that change, the model can be written

$$r = \frac{w}{4\pi} \frac{\mu_0}{\mu + \mu_0} \{P(\alpha) + H(\mu)H(\mu_0) - 1\} \qquad 5.85$$

The H-functions are technically known as the Ambartsumian–Chandrasekhar H-functions for the two astronomers, **Victor Ambartsumian** (1908–1996) and **Subrahmanyan Chandrasekhar** (1910–1995), who derived mathematical methods for determining the role of multiple scattering in the radiative transfer problem. Their exact solution requires numerical methods, but

5.7 More Complex Photometric Models

Hapke (1981) provides an approximation that is accurate everywhere to within a few percent:

$$H(x) = \frac{1 + 2x}{1 + 2x\sqrt{1 - w}} \qquad 5.86$$

Technically, Equation 5.85 is only exact for the isotropic scattering case ($P(\alpha) = 1$), but Hapke (1981) argues that, if the scattering function is not too anisotropic, multiple scattering will tend to randomize any directional tendencies, making it a reasonable approximation. As noted in Chapter 4, this is referred to as the *isotropic multiple-scattering approximation* or IMSA.

This model was subsequently modified by Hapke (2002) to take into account non-isotropic multiple scattering. It is considerably more complex than the IMSA and so we will not discuss it here. The interested reader is encouraged to consult that reference for more details.

5.7.1.2 Opposition Surge

Hapke (1986) incorporated the opposition surge to due shadow hiding (SHOE, Chapter 4) into his model as

$$r = \frac{w}{4\pi} \frac{\mu_0}{\mu + \mu_0} \{P(\alpha)[1 + B(\alpha)] + H(\mu)H(\mu_0) - 1\} \qquad 5.87$$

where

$$B(\alpha) = \frac{B_0}{1 + \left(\frac{1}{h}\right)\tan\frac{\alpha}{2}} \qquad 5.88$$

B_0 is the **amplitude of the opposition surge** – a factor that tells us how much brighter the reflectance can get above the expected value (i.e. with no surge) when $\alpha = 0°$. The maximum theoretical value for B_0 is 1.0 and the maximum value for $B(0°)$ is also 1.0, meaning that the opposition surge can cause a doubling in the single-scattering intensity. However, when applied to actual spacecraft or telescopic data it has been found that, when allowed to float in a fitting algorithm, B_0 is often >1.0. Hapke (1993) offers possible reasons for this, including a glory for spherical particles (considered unlikely in regoliths), corner reflections from crystal faces (also considered unlikely in regoliths continuously bombarded by micrometeorites), and an intrinsic shadow hiding for the complex shaped particles making up the regolith. This latter explanation may be viable and should rightly be included in the particle phase function; however, most models of the particle phase function (e.g. Henyey–Greenstein, Chapter 4) do not have the degrees of freedom necessary to accurately model an opposition surge on top of an otherwise smooth particle phase function, leading to an artificially high B_0 in compensation.

A measure of the width of the opposition surge is given by h; in practice, the angular half-width (in radians) of the opposition surge is well approximated by $2h$. Hapke (1986) estimates several values for h given different assumptions about the particle size distribution; with realistic distributions, values of 0.05–0.09 are commonly expected and observed, corresponding to angular half-widths of 6°–10°.

5.7.1.3 Surface Roughness

Hapke (1984) assumes the surface can be broken into facets that are oriented randomly in azimuth and obey some slope distribution function. Each facet scatters according to the Hapke function assumed for smooth surfaces, and the mean incidence and emission angle are modified because each facet has a new modified orientation with respect to the source and viewer. Hapke's shadowing function contains a single **roughness parameter**, $\bar{\theta}$, (theta-bar) which is a measure of the surface roughness. Theoretically, it is an average of all the slopes ranging from the footprint of the detector in scale down to that of individual particles. A physical interpretation has been more difficult to specify (Helfenstein, 1988; Shepard and Campbell, 1998).

Hapke notes that his shadowing function, $S(\bar{\theta})$, only accounts for single scattering. Multiple scattering can reduce the effects of shadowing (by illuminating otherwise dark areas), and so the roughness parameters should be considered a lower limit for brighter surfaces. It is incorporated into the overall Hapke model as a multiplicative factor:

$$r = \frac{w}{4\pi} \frac{\mu_{0e}}{\mu_e + \mu_{0e}} \{P(\alpha)[1 + B(\alpha)] + H(\mu_e)H(\mu_{0e}) - 1\} S(i, e, \alpha, \bar{\theta}) \qquad 5.89$$

The incidence and emission angle terms are now *effective* incidence and emission (subscript e) because the assumption of tilted facets means that each facet has a slightly different incidence and emission angle, and the effects of this must be averaged. The shadowing function itself is analytical, but elaborate, with versions that depend upon whether $i > e$ or vice versa; we will not reproduce it here. The interested reader is referred to Hapke (1984) or Hapke (1993, 2012). It is important to note that $0 \leq S \leq 1$; in other words, shadowing can reduce the expected reflected light but not increase it as the opposition surge does.

5.7.1.4 Other Modifications

The latest version of the Hapke model includes more recently discovered effects. For example, it now includes an opposition effect created by coherent backscattering (Chapter 4; Hapke, 1990). There is also a modification to better account for the role of anisotropic phase functions in multiple scattering (the H-functions). And most recently, Hapke (2008) incorporated the effects of finite particle size and porosity after laboratory studies demonstrated that variations in porosity cause large variations in the brightness of surfaces. All of these add additional free parameters and layers of complexity, and the interested reader should consult those references for additional details.

5.7.1.5 Disk-Integrated Form

Earlier, we derived the disk-integrated photometric properties and functions of simple scattering laws: Lambertian, Lommel–Seeliger, and the Area Law. These have also been derived for the Hapke model (see Hapke, 2012), and we list them here without derivation. The functions, as listed here, assume no roughness parameter and no coherent-backscatter opposition effect, but do include multiple scattering and the shadow-hiding opposition effect.

In Chapter 4, we introduced two auxiliary functions that Hapke (2012) makes use of in his solution to the equation of radiative transfer: the *albedo factor*, γ, given by

$$\gamma = \sqrt{1-w} \qquad 5.90$$

and the so-called *diffusive reflectance*, given by

$$r_0 = \frac{1-\gamma}{1+\gamma} \qquad 5.91$$

For a surface conforming to a Hapke scattering model, the geometric albedo is well approximated (better than 0.4% everywhere) by (Hapke, 1993, 2012)

$$p_v = \left(0.49 r_0 + 0.196 r_0^2\right) + \frac{w}{8}[(1+B_0)P(0) - 1] \qquad 5.92$$

The second term of this equation handles non-isotropic scattering and the opposition surge. If we ignore the opposition surge ($B_0 = 0$) and consider only isotropic scattering ($P(0) = 1.0$), it reduces to

$$p_v = 0.49 r_0 + 0.196 r_0^2 \qquad 5.93$$

The integral phase function is given by

$$\Phi(\alpha) = \frac{r_0}{2 p_v} \{\Phi_1 + \Phi_2\} \qquad 5.94$$

where Φ_1 has the form of the integral phase function of a Lommel–Seeliger surface modified by a non-isotropic phase function and shadow-hiding opposition effect, given by

$$\Phi_1(\alpha) = \left[1 + \sin\frac{\alpha}{2}\tan\frac{\alpha}{2}\ln\left(\tan\frac{\alpha}{4}\right)\right]$$
$$\times \left[\frac{(1+\gamma)^2}{4}\{[1+B(\alpha)]P(0)-1\} + (1-r_0)\right] \quad 5.95$$

If we ignore the opposition surge ($B(\alpha) = 0$) and assume isotropic scattering ($P(0) = 1$), this reduces to

$$\Phi_1(\alpha) = (1-r_0)\left[1 + \sin\frac{\alpha}{2}\tan\frac{\alpha}{2}\ln\left(\tan\frac{\alpha}{4}\right)\right] \quad 5.96$$

As w approaches 0, this approaches the Lommel–Seeliger integral phase function. The second term, Φ_2, is the part of the integral phase function, which incorporates multiple scattering (the H-functions) and is similar in form to that of the Lambertian integral phase function:

$$\Phi_2(\alpha) = \frac{4}{3\pi}r_0[\sin\alpha + (\pi - \alpha)\cos\alpha] \quad 5.97$$

Taken together, we see that the Hapke model integral phase function is similar to that of the Lunar–Lambert model. Finally, the disk-integrated Bond albedo of a surface obeying the Hapke model is given (after some rearrangement of Hapke's (2012) original formulation) by

$$A = \frac{1}{6}\left(5r_0 + r_0^2\right) \quad 5.98$$

5.7.2 Lumme–Bowell

In 1981, Hapke published the first outline of his model for optical scattering from planetary objects without atmospheres. In an interesting coincidence of timing, two other groups published on the same problem at the same time. **Kari Lumme** (University of Massachusetts) and **Edward Bowell** (Lowell Observatory) published a similar radiative-transfer-based model, and **Jay Goguen** (JPL) completed his Ph.D. thesis at Cornell on this problem (with similar solutions), all in 1981. Lumme and Bowell's (1981) model was an early competitor to the Hapke model, but never caught on in the same way as the Hapke model and eventually fell out of use as a comprehensive analysis tool. However, it played a fundamental role in the development of the modern asteroid phase curve system (the HG system, Chapter 6).

Nice summaries of the Lumme–Bowell model are given by Karttunen (1989) and Fairbairn (2004). The bidirectional reflectance of their model can be written

$$r = \frac{w}{4\pi} \frac{\mu_0}{\mu_0 + \mu} (2\Phi_S + \Phi_M) \qquad 5.99$$

where Φ_S is the phase function for single scattering and Φ_M is that for multiple scattering[3]. Right away, we notice similarities to the Hapke model in that it is based on the principles of radiative transfer and the first term is the Lommel–Seeliger function.

The single-scattering phase function is a product of three different functions

$$\Phi_S(\alpha) = P(\alpha) \Phi_{oppsurge}(\alpha, D) \Phi_{roughness}(\alpha, q, \rho) \qquad 5.100$$

where $P(\alpha)$ is the single particle phase function, often given by a one-term Henyey–Greenstein function (Chapter 4), and the next two functions account for the opposition surge and surface roughness.

The opposition surge function is given by

$$\Phi_{oppsurge} = \exp\left\{ \frac{-\sin\alpha}{0.636 D + 1.828 \sin\alpha} \right\} \qquad 5.101$$

where $0 \leq D \leq 1$ is the volume density (1 – *porosity*). We note that in the original Lumme and Bowell (1981) paper as well as the two cited earlier, this parameter is referred to as the "phase function due to shadowing." Here we have renamed it to indicate that it accounts for the opposition surge to avoid confusing it with Hapke's earlier terminology where shadowing function refers to the effects of topographic roughness. Lumme and Bowell (1981) also emphasize the volume density is a *photometric* concept that may or may not have a direct physical interpretation.

The roughness phase function component is given by

$$\Phi_{roughness} = \frac{1 + (1-\varepsilon) v}{1 + \rho \varepsilon v} \qquad 5.102$$

where $0 \leq \varepsilon \leq 1$ is the fractional area of the surface covered by "holes," basically any depression from any cause at any scale from clumps of particles upward, and $\rho \geq 0$ is the ratio of mean hole depth-to-radius – a measure of roughness. To first order, the relationship between the Lumme–Bowell and

---

[3] This is an example of an unfortunate overlap in symbols common in the literature. Here, the Greek capital phi (Φ) is a scattering phase function for disk-resolved observations, NOT the integral phase of Section 5.5.3. However, this notation is used in the references listed, and we follow it here.

Hapke roughness parameters is $\rho = \tan(\bar{\theta})$. The final variable in this parameter is given by

$$v = \frac{(\mu^2 + \mu_0^2 - 2\mu\mu_0 \cos\alpha)^{\frac{1}{2}}}{\mu\mu_0} \qquad 5.103$$

The numerator of this odd-looking function is the relative azimuth between the incidence and emission vectors (derived from the law of cosines) and the entire function can be thought of as taking into account the overlap between cast and viewing shadows.

The multiple-scattering phase function, Φ_M is identical to that used by Hapke (1981):

$$\Phi_M(\alpha) = H(\mu)H(\mu_0) - 1 \qquad 5.104$$

Like the Hapke model, the Lumme–Bowell model assumes that multiple scattering can be approximated by an isotropic particle phase function while allowing the single-scattering term to enjoy anisotropic behavior.

Although the Hapke and Lumme–Bowell models have different ways of accounting for the opposition surge and topographic roughness, they are structurally similar. If we take *only the single-scattering* components of both models and reorganize them slightly, we get

$$\underset{\text{Hapke}}{\frac{w}{4\pi}\frac{\mu_0}{\mu+\mu_0}\{P(\alpha)[1+B(\alpha)]\}S} = \underset{\text{Lumme–Bowell}}{\frac{w}{4\pi}\frac{\mu_0}{\mu_0+\mu}2P(\alpha)\Phi_{oppsurge}\Phi_{roughness}}$$

or

$$P(\alpha)S + P(\alpha)B(\alpha)S = 2\,P(\alpha)\Phi_{oppsurge}\Phi_{roughness} \qquad 5.105$$

and can see the similarities (we have ignored the slight changes in the effective incidence/emission angle in the Hapke roughness correction). Recall that $B(\alpha)$ is the Hapke opposition surge parameter and S is the Hapke roughness correction factor. Both models assume that the opposition surge is strictly a single-scattering phenomenon, but the Lumme–Bowell model counts the opposition surge term twice compared with Hapke.

The multiple-scattering terms of both models are identical

$$\frac{w}{4\pi}\frac{\mu_0}{\mu+\mu_0}\{H(\mu)H(\mu_0) - 1\}S(i,e,\alpha,\bar{\theta}) = \frac{w}{4\pi}\frac{\mu_0}{\mu_0+\mu}(H(\mu)H(\mu_0) - 1)$$

except that Hapke also applies the topographic roughness correction to this term, while Lumme–Bowell does not.

In Chapter 6, we will show how this model forms the foundation of the modern system for describing the absolute magnitude and phase curve behavior of asteroids.

5.8 Extracting Parameters from Complex Models

Although tangent to the main topic of this chapter, it is important to briefly discuss some of the methods used to extract model parameters from a set of observations. For the simpler models described earlier, it is relatively straightforward to find the best-fit reflectance parameters (e.g. single-scattering albedo). It is a different matter to find the best-fit parameters from more complex models, such as the Hapke model, and it is possible that there may not be a unique solution.

At a minimum, the Hapke model involves four parameters: single-scattering albedo, w, opposition surge width, h, and magnitude, B_0, and a one-term particle phase function. Most studies use at least a two-term particle phase function to maximize the model flexibility, bringing the typical parameter total to five. In many studies, the surface roughness parameter, theta-bar, is also included, raising the total to six parameters. In more recent incarnations, the Hapke model can have an additional parameter for porosity, two terms for a coherent backscatter opposition effect contribution, and an optional third term to describe the particle phase function.

The difficulty is that most of the Hapke parameters change the scattering function in non-linear ways. The solution space is large, and finding a true minimum is not easy. Below, we describe a few of the methods commonly used to find the best fit as characterized by the chi-squared parameter:

$$\chi^2 = \sum \frac{(r_{obs} - r_{mod})^2}{\sigma^2} \qquad 5.106$$

where r is the reflectance term (e.g. radiance factor), the subscripts indicate the observed (*obs*) value and the model (*mod*) prediction for a given set of conditions, and the denominator is the variance in the observations.

Determining the uncertainties in retrieved model parameters is non-trivial. For some parameters, small variations may have a large effect on chi-squared, i.e. they are tightly constrained, but other parameters may be so unconstrained by the observations that they have little effect on the final fit. A discussion of these issues is beyond the scope of this book, but one should expect to find some discussion of the methods used to determine uncertainty in any photometric study.

5.8.1 Grid-Search Method

A grid-search method is exactly as it sounds. All parameters are varied in a uniform and systematic way to search the entire sample space. For example, one may test solutions in which the single-scattering albedo varies from 0.0 to 1.0 in increments of 0.02 units, the surge width parameter varies from 0.0 to 0.3 in increments of 0.01 units, etc. Depending on the coarseness of the grid and the number of parameters to search on, this can take a significant amount of time, even with a computer. A six-parameter model divided into increments of 50 divisions each requires almost 16 billion tests, and even so, there is no guarantee that one of these will be the true minimum.

Obviously, this method can be optimized with *a priori* knowledge; for example, there is no need to test single-scattering albedos up to 1.0 for the darker objects in the solar system. And a common method employed to speed the process is to do a coarse search, find one or more of the best solutions, then do finer grid searches only around these areas.

5.8.2 Gradient-Search Method

In this method, one inputs a series of **seed values** for the model parameters. The routine calculates the chi-square, and then starting with one parameter (e.g. single-scattering albedo) changes the value by a small increment in one direction, re-calculates, and compares the new chi-square. If it is smaller, it continues changing the parameter in that direction until the chi-square stops decreasing. (If chi-square increases after the first increment in one direction, the routine will check by incrementing in the other direction.) Once the chi-square stops decreasing, the routine will move to the next parameter (e.g. opposition surge width) and repeat this process. Once all parameters have been tested, it repeats the cycle until it cannot get the chi-square to drop more than a small fractional amount, called the **threshold,** which is set by the user at the beginning of the search. If it is too small, the search may become stuck in a long and unproductive loop. If it is too large, it may not search long enough.

Even after a gradient-search method has stopped with a best-fit answer, it may not be the best one. In a large sample space with multiple parameters, there are likely to be many local minima. If the algorithm reaches one of these, it will stop looking because there are no solutions with smaller chi-square in the vicinity. The only way to check is to run the search again using different seed values. To have confidence in a gradient-search solution, one must do many searches using many different seed values. In practice, many users run a combination of grid- and gradient-search methods to increase confidence in the reported best-fit parameters.

5.8.3 Genetic Algorithms

Like organisms in an ecosystem, genetic algorithms use a survival-of-the-fittest approach to find the best solutions. Although still uncommon in planetary photometry, they have been used in a few studies. An example of one implementation is given in Cord et al. (2003) as they attempted to fit laboratory data with a six-term Hapke model. A **chromosome** was defined as a set of six Hapke parameters, and they began the search with 10,000 randomly selected chromosomes. Each chromosome was tested and the best were selected as **parents**; these were allowed to cross-pollinate with each other to produce new chromosomes, **children**, whose attributes were a mix of those from each parent with an additional element of mutation (randomness). Those children with the lowest chi-square were allowed to live and reproduce; the others were eliminated. The process was repeated until, several generations of chromosomes later, a threshold for goodness-of-fit was reached and the search stopped.

As with any search algorithm, there is a large element of empirical tweaking that must be done for it to work well. Questions to address include how many initial chromosomes are needed? How are they to be cross-pollinated to form children? How much mutation is used, and what criteria are used for selecting the best?

6
Planetary Disk-Integrated Photometry

Much of the work in photometry involves making accurate measurements of several planetary quantities: the magnitude (brightness), color, polarization, and variations in these quantities with time and phase angle. Those data are analyzed to learn something about the object in question, and to characterize its brightness so that we can predict what it will be under any illumination and observation conditions. In this chapter, we examine disk-integrated observations – those generally made by ground-based telescopes.

There are at least four planetary characteristics that can often be extracted or constrained with disk-integrated observations: (1) rotation rate, (2) size, (3) intrinsic reflectivity – an indicator of underlying composition, and (4) some properties of the visible surface, such as whether it's gaseous or particulate. For smaller objects of non-spherical shape, we can also place some constraints on their aspect ratio and, occasionally, gross shape.

In the first section, we show how the phase curve and extracted absolute magnitude of an object can be used to estimate a planetary albedo and diameter. We look at the lightcurve, the time variation of an object's brightness as it rotates and the influence of albedo variations and an irregular shape. The phase curves of asteroids have their own nomenclature that requires some explanation and we examine other phase effects, such as the observed polarization and spectral color of objects. Finally, we conclude with a case study – published examples of the application of this material to Mercury.

6.1 Planetary Size and Albedo

6.1.1 Derivation

If we have a suite of planetary observations and know the orbit, we can find the reduced magnitude for every observation using Equation 2.23. With enough of

6.1 Planetary Size and Albedo

these observations, we can use one of the methods described in Chapter 2 to estimate an absolute magnitude. And with the absolute magnitude, we can constrain the size and albedo of the planet as follows.

The geometric albedo is defined as the ratio of the visual light flux (measured power, irradiance, or radiant intensity) from an object at $\alpha = 0°$ and distance from Earth, Δ, to that measured from a perfect Lambertian disk of the same cross-sectional area at the same distance and phase angle. Although there are several ways to proceed, I find it easier to understand by writing everything in terms of the power measured by the detector of sensor area, dA. We can then write the geometric albedo as

$$p_v = \frac{E_{obj}(0,\Delta)dA}{E_{Lam}(0,\Delta)dA} \qquad 6.1$$

The power measured at the detector from a Lambertian surface (disk) of cross-sectional area equal to the planet is

$$E_{Lam}(0,\Delta)dA = \left(\frac{E_{inc}}{\pi}\right)(\pi R^2)\left(\frac{dA}{\Delta^2}\right) = E_{inc}\left(\frac{R}{\Delta}\right)^2 dA \qquad 6.2$$

where the terms of the intermediate expression have been grouped for ease of understanding; the first term is the radiance reflected from an illuminated perfect Lambertian surface (incident irradiance E_{inc}) at zero phase angle, the second term is its area assuming it to be a disk of the same radius, R, as the planet, and the third term is the solid angle of the detector a distance Δ away. The final term in parentheses is sometimes referred to as the *area factor*.[1]

If we substitute Equation 6.2 into 6.1, we get the following:

$$p_v = \frac{E_{obj}(0,\Delta)}{E_{inc}}\left(\frac{\Delta}{R}\right)^2 \qquad 6.3$$

If the object is 1 AU from Earth and the Sun and assuming visual wavelengths, we can rewrite this in terms of its absolute visual magnitude and that of the Sun (see Eq. 2.20) as

$$p_v = 10^{-0.4[V(0)-V_{Sun}]}\left(\frac{1\ AU}{R}\right)^2 \qquad 6.4$$

If we take the log of both sides of Equation 6.4, we get

$$\log p_v = 0.4[V_{Sun} - V(0)] + 2\log(1\ AU) - 2\log R \qquad 6.5$$

---

[1] This term is often written as $\sin^2(R/\Delta)$ for reasons unknown to us, but for small R/Δ (any realistic planetary application) these are indistinguishable.

and if we convert the object's radius into diameter and regroup terms, we get

$$\log p_v = (0.4 V_{sun} + 2\log(1\text{ AU}) + 2\log 2) - 0.4 V(0) - 2\log D \quad 6.6$$

Assuming the solar magnitude to be -26.75 (Cox, 2000) and summing the terms in parentheses (length and diameter terms are in kilometers) leads to the commonly used equation[2]:

$$\log p_v = 6.252 - 0.4 V(0) - 2\log D \quad 6.7$$

or

$$D = \frac{1336}{\sqrt{p_v}} 10^{-0.2 V(0)}$$

6.1.2 Asteroids and Comets

6.1.2.1 Asteroids

In practice, the formulae derived above for spherical objects hold for the smaller asteroids, with two caveats. Small bodies are irregular and their brightness varies on the order of hours as they rotate. To account for this, we approximate using the mean absolute magnitude (average over a complete lightcurve rotation – see Section 6.2), and we replace the diameter with an **effective diameter**, D_{eff}, the diameter of a sphere with the same volume as the irregular asteroid.

6.1.2.2 Comets

There are several issues that make it more difficult to characterize the brightness and physical properties of comets. They are dynamic, almost by definition; their size and albedo often change as they orbit the Sun. The **nucleus** is the heart of the comet and chiefly all that is seen when far from the Sun. Composed largely of ice and silicates, it sheds dust and gas as it warms on the approach to the Sun, creating an extended **coma**, which often completely hides the nucleus from view. Because of these observational differences with asteroids, comets are often assigned at least two apparent magnitudes: one for the nucleus of the comet, called the **nuclear magnitude**, m_{nuc}, and one for the nucleus plus the extensive coma consisting of gas and dust, called the **total magnitude**, m_{tot}.

The nuclear magnitude can be determined only when the comet is far from the Sun and inactive. In that case, one can use the standard magnitude versus

---

[2] Our values for these constants disagree slightly with others in the literature. Where we have 6.252 and 1,336, older references give 6.247 and 1,329. These are likely due to slight differences in the assumed visual magnitude of the Sun.

inverse square formula to the data and attempt to constrain the diameter and albedo using Equation 6.7. That equation can also be used to test the assumption that the comet is inactive since it will only be obeyed if there is no additional coma contribution to the nuclear brightness.

The total magnitude is of more interest to comet researchers because it can be used to estimate the activity of the comet. However, its measurement is fraught with peril. The coma is an extended source, unlike the nucleus, and requires an area measurement on the image frame. To be accurate, this requires significantly more work in removing systematic biases in the image frame (flat-fielding), ensuring the measurement radius captures the entire coma, and removing the inevitable contribution of other stars within the coma field of view.

Perhaps more problematically, the brightness of the coma is a composite of reflection from dust and icy particulates – a continuous reflectance spectrum called the **continuum**, and a variety of emission lines superimposed on the continuum. This **emission spectrum** is due chiefly to photoionization and resonance fluorescence of a variety of gaseous species released as the nucleus warms (Wyckoff, 1983; also see Chapter 2). The ratio of gas to dust is a quantity that has implications for the origin and age of the comet, and one that changes with heliocentric distance, even varying between the approach to and departure from perihelion (Marsden and Roemer, 1983; Meisel and Morris, 1983).

To best characterize the species of volatiles making up the nucleus and to estimate production rates of gas and dust, comets are often observed in narrowband filters that isolate the emission spectrum of specific components, usually daughter products created by the dissociation of compounds in the nucleus. For example, the CN molecule has an emission band at 0.3875 μm (Schleicher and Farnham, 2004). In Chapter 2, we briefly reviewed the primary photometric systems used by comet researchers and noted that many filters were chosen to characterize five or six of the most common emission lines. Other filters are chosen to avoid these emission lines and get a better estimate of the background or continuum reflectance properties of the coma.

Assuming that one correctly measures the total apparent magnitude of the comet in a specific wavelength band, its relationship to orbital distance is often written as (Marsden and Roemer, 1983)

$$m_{tot}(\alpha, d, \Delta) = m_{tot}(1, \alpha) + 2.5 \, c_1 \log d + 2.5 \, c_2 \log \Delta \qquad 6.8$$

where c_1 and c_2 are parameters that account for the active change in brightness with changing heliocentric distance, and again, the symbol V would be substituted for m in the specific case of visual magnitudes. In an ideal case, a number of observations of the comet are acquired and the parameters are solved for in a least squares sense.

In the absence of multiple observations, researchers are often forced to choose values for c_1 and c_2. Choosing $c_1 = c_2 = 2$ will make this equation into an inverse square law, but it is more common for investigators to choose $c_1 = 4$ and $c_2 = 2$ to account for the rapid brightening of the nucleus with decreasing heliocentric distance. One must be careful when using brightness values from older papers; in many, it was common to refer to the reduced visual magnitude $V_{tot}(1, \alpha)$ as H_0 or H_1 (cf. Marsden and Roemer, 1983; Meisel and Morris, 1983), which is unfortunately similar to the modern symbol for absolute magnitude. When $c_1 = 4$ the term in front of $\log(d)$ becomes 10, and the reduced magnitude is then commonly referred to as the H_{10} magnitude (Vsekhsvyatskii, 1964).

6.2 Lightcurves

All solar system objects vary in brightness with time as they rotate, but the reasons vary. For some objects, it is because of albedo variations on the surface; for others, it is because their shape is irregular. For a few objects, both factors may come into play. For active comets, their brightness is a function of their volatile content and distance from the Sun.

6.2.1 The Shapes of Things

In 2006, the International Astronomical Union formally defined the term **planet** after the discovery of the trans-Neptunian object 136199 Eris. Early estimates of its size indicated that it was comparable to, if not larger than Pluto. Thus, the great planetary debate ensued – should Eris be a planet? Or should Pluto be downgraded into a new status, yet to be defined? The down-graders carried the debate, and Pluto, Eris, and the largest asteroid 1 Ceres were defined to be **dwarf planets**. Anything that didn't fit the definition of planet or dwarf planet was lumped into the catch-all term *Small Solar System Body*.

The definition of both planet and dwarf planet include the following condition: it must have "sufficient mass for its self-gravity to overcome rigid body forces so that it assumes a hydrostatic equilibrium (nearly round) shape" (Resolution 5, in IAU, 2008). Essentially, when a planetary object reaches a certain mass, its own self-gravity is stronger than the material strength of the ice, rock, and metal that make it up, forcing it into a spherical shape. If the object is rotating rapidly, centrifugal forces can drive the shape from spherical into something more ellipsoidal, but this is considered hydrostatic equilibrium and so still fits the definition.

Where is this mass tipping point that separates objects that can be irregular and those that must be nearly spherical? It is a fuzzy boundary that depends upon the materials in question, but in our own Solar System, asteroid 1 Ceres is 950 km in diameter, spherical, and over the dwarf-planet line. The second largest asteroid, both in terms of mass and diameter, 4 Vesta, is 525 km in diameter, and although "roundish," is not in hydrostatic equilibrium and therefore under the dwarf-planet line. All other asteroids, including the Trojans, are smaller than Vesta and, as a starting point, can be assumed to be irregular in shape. A significant number of the known trans-Neptunian objects are larger than Vesta and must be assumed to be in hydrostatic equilibrium.

6.2.2 Lightcurves of Simple Shapes

6.2.2.1 Shape Effects Only

Imagine a homogeneous sphere orbiting the Sun and spinning on a **principal** (single) **axis** pointed in any direction. Because neither the visible albedo nor apparent cross-sectional area changes, it will show no variations in reflected light as it spins. However, if it is elongated or distorted in any way, its brightness will vary periodically.

Consider the simplest non-spherical case – an ellipsoid defined by three axes: a, the semi-major axis; c, the semi-minor axis and axis of rotation; and b, an intermediate axis perpendicular to both with dimensions $c \leq b \leq a$. The **effective diameter**, D_{eff}, of a sphere with this volume would be

$$D_{\mathit{eff}} = 2\,(abc)^{\frac{1}{3}} \qquad 6.9$$

Figure 6.1 shows a schematic of the lightcurve of such an object and its appearance for two different orientations of its principal axis. In the first case, we see the object from an **equatorial aspect**. In this case, we see the greatest variation in reflected light as the apparent surface area increases and decreases with rotation. It is common to refer to the **aspect** in terms of the **sub-Earth latitude**, the latitude on the surface where a line from the observer on Earth intersects the object's surface before continuing to the center of the object.

If, for ease of computation, we assume an object obeys the Area Law (Chapter 5), its brightness varies linearly with the visible area. For an ellipsoid, the cross-sectional area is $A = \pi xy$ where x is the *apparent* semi-major axis and y the *apparent* semi-minor axis. For an ellipsoid obeying the Area Law, its reflectance, r, can thus be written $r = kxy$ where k is a constant that takes into account the π and reflectance of the surface material. If the b and c axes of our ellipsoid are equal, the brightness at maximum is kac, and at minimum is kc^2. The brightness of the asteroid is thus a/c times higher at maximum than at minimum so that the **amplitude** of the lightcurve is equivalent to the a/c axis ratio. This is an oft-used approximation for estimating the minimum aspect ratio of an asteroid.

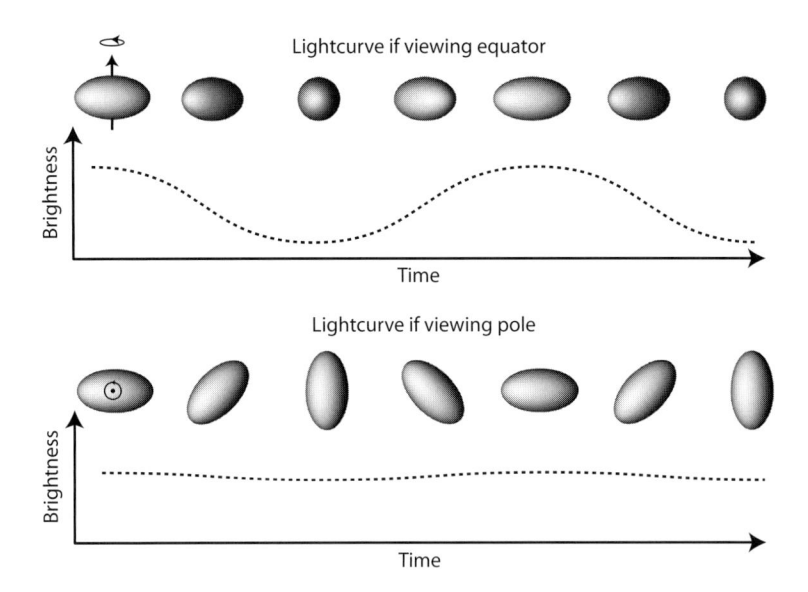

Figure 6.1. The top figure shows the lightcurve of an ellipsoid-shaped object if viewed from an equatorial aspect, while the bottom shows if viewed from a near-polar aspect.
Credit: Copyright M. Shepard, Asteroids: Relics of Ancient Time, Cambridge University Press. Reprinted with permission.

In the second case shown, we see the object from a near **polar aspect** and, as with the homogeneous sphere, there is little or no variation in apparent brightness with rotation – we always see the same cross-sectional area. If seen from a mid-latitude aspect, the amplitude of the lightcurve will be intermediate to the other two cases.

The shape of the lightcurve in this simplified case is sinusoidal and one complete rotation includes two identical maxima and two identical minima. The maxima represent views of the "front" and "back" sides while the minima are the two end-on views. The time interval between these is the **rotation period**. Even with more complex shapes, lightcurves of most small objects have two maxima and two minima, but in general they will not be identical. These waveforms are referred to as **bimodal**, meaning two maxima, and by implication two minima.

6.2.2.2 Albedo Effects Only

A spherical asteroid with an albedo difference in longitude would display a lightcurve with only a single maxima and minima as it rotated – a unimodal lightcurve. But how would you know this was the case? Two rotations of this object would have a bimodal lightcurve and be difficult to distinguish from a single rotation of a uniform albedo ellipsoid. Unless one has independent evidence for a rotation period, perhaps from radar observations or spectral variations synchronous with the lightcurve, it may be impossible to tell these cases apart.

6.2.3 Lightcurves of Irregular Shapes

It is a rare small body that has a simple shape such as a sphere or ellipsoid, and more complex shapes give rise to more complex lightcurves. This is actually good news for the researcher because it makes it easier to refine the rotation period by using small irregularities in the lightcurve. However, it also requires something more sophisticated than a simple sine function to characterize the lightcurve. For this, we turn to **Fourier analysis**.

A **Fourier series**, named for the physicist **Joseph Fourier** (1768–1830), is a sum of sine and cosine functions that allow one to describe a more complex periodic wave form. For a periodic lightcurve of brightness f at time t, we can characterize its wave form as

$$f(t) = a_0 + \sum_{n=1}^{\infty} \left[a_n \cos\left(\frac{2\pi nt}{T}\right) + b_n \sin\left(\frac{2\pi nt}{T}\right) \right] \qquad 6.10$$

Here, T is the rotation period of the object, and a_n and b_n are amplitude coefficients that give greater or lesser weight to sine or cosine functions of different integral periods n/T. The coefficient a_0 gives a baseline or mean brightness and is often set to $a_0 = 0$ when we are only concerned with finding the period. The number of terms, n, is often referred to as the **harmonic order** of the series.

To find the value of coefficients necessary to reproduce $f(t)$, one can use the **Euler formulas** (Kreyszig, 1983), named for the mathematician **Leonhard Euler** (1707–1783).

$$a_0 = \frac{1}{T} \int_{-T/2}^{T/2} f(t)\,dt$$

$$a_n = \frac{2}{T} \int_{-T/2}^{T/2} f(t) \cos\left(\frac{2\pi nt}{T}\right) dt \qquad 6.11$$

$$b_n = \frac{2}{T} \int_{-T/2}^{T/2} f(t) \sin\left(\frac{2\pi nt}{T}\right) dt$$

In most asteroid lightcurve applications, uncertainty in the data makes it unnecessary to use more than ten harmonic orders in the series, and often only half that many (Warner, 2006). These functions are also commonly written using complex number notation, but we will not reproduce them here.

A rare few asteroids show multi-modal – three and four maxima – per rotation. What is the cause of these oddities? It is likely that the object is quite irregular and shadowing plays a big role, especially at higher phase angles. There may also be some combination of irregularity and albedo variations.

6.2.4 Asteroid Shapes from Lightcurves

Until spacecraft visited the first asteroids, there was no consensus on their surface condition. What were their shapes? Did they have regolith? Were there regions of brighter and darker material as we see on the Moon and many other objects? Shortly after their discovery in the early nineteenth century, astronomers noted that asteroids regularly varied in brightness, and many, believing them to be fragments of a now destroyed planet, attributed these variations to an irregular shape, a remarkably prescient interpretation (Shepard, 2015).

It is a short step from that to wonder whether, given enough lightcurves at different aspects, it might be possible to use them to estimate the asteroid's shape. This dream was temporarily crushed by a single sentence from Henry Norris Russell. In his paper entitled *On the light-variations of asteroids and satellites* (Russell, 1906), he concludes the following:

> It is quite impossible to determine the shape of an asteroid. If any continuous convex form is possible, all such forms are possible.

Ouch. However, Norris – not knowing anything of the true nature of asteroids – considered variations of both shape and albedo. For this general case, his statement holds. There are just too many free variables if both the shape and the distribution of albedo spots is arbitrary. However, if the albedo is constant, then only the shape need be considered. But Norris declined to prosecute this case stating

> But this leads to difficult problems in the theory of surfaces, and will not be attempted here.

Norris had no reason to believe that asteroids would be any different from planets in showing wide changes in terrain and albedo on the surface. But in the past few decades, we have visited a dozen or so asteroids and half that many comet nuclei with spacecraft, and a relevant trend has been noted: almost all of the smaller asteroids and (inactive) comet nuclei observed by spacecraft have shown very little variation in albedo across their surface. The only exceptions to this are the dwarf planet 1 Ceres and the second largest asteroid

4 Vesta, both of which have shown subtle global albedo variations in telescopes and more significant variations at spacecraft scales (Gaffey, 1997; Reddy et al., 2013; Li et al., 2016).

In 1988, **Steven Ostro** (1946–2008) and colleagues (Ostro et al., 1988) revisited the problem and determined that, with some assumptions, including that the surface albedo is homogenous, a convex shape can be determined from a series of lightcurves. This underlying assumption has been borne out by numerous spacecraft observations. In 1992, **Mikko Kaasalainen** (Tampere University of Technology) and colleagues (1992a, b) made additional progress on the problem and, a decade later, published methods for optimizing this inverse problem with illustrated examples (Kaasalainen and Torppa, 2001; Kaasalainen et al., 2002). There can be no asperities cast on the reluctance of Dr. Russell to tackle the problem since it has only become possible to solve with the advent of the modern desktop computer.

In general, the shape solution requires lightcurve observations of an asteroid at a wide variety of aspects, the assumption of a uniform surface albedo, and an assumed light-scattering law, and is restricted to providing convex shape solutions. A convex shape is one that has no indentations, and is often described as the shape one would get if the asteroid were loosely gift-wrapped (not shrink-wrapped) so that there is space between any concavities and the final shape. Experiments suggest that, with these restrictions, there is one unique convex shape that will fit a series of lightcurves (Kaasalainen and Torppa, 2001). The inability to find such a convex shape is one sign of possible albedo variations.

In recent years, this work has been applied by Durech and colleagues (2010) to generate and maintain a large online database of asteroid shapes, rotation periods, and pole orientations. This database also contains notes when fits suggest albedo variations or bifurcated (contact binary) shapes.

6.2.5 Comet Lightcurves

An important point of terminology is that when comet scientists refer to a comet's **secular light curve**, they are referring to the brightness behavior of a comet as it orbits the Sun, *not* its rotational lightcurve (Ferrin, 2010). In this section, we use the term lightcurve to mean that due strictly to *the rotation of the nucleus*.

Comets present special problems when attempting to obtain and analyze a rotational lightcurve. First, the nucleus, and its associated rotation period, is only visible when the comet is several AU from the Sun and inactive; closer to the Sun, we only see the coma. Second, a comet's rotation state is not always well behaved. Most asteroids are in what is referred to as a **principle axis rotation state** – they rotate as a rigid body about a fixed axis. These spin states are usually stable over thousands of years, only changing because of collisions, close approaches to planets, or sunlight-driven forces such as the Yarkovsky-O'Keefe-Radzievskii-Paddack or YORP effect (Bottke et al., 2006). While a few asteroids have been discovered that do not rotate in this way, i.e. they are "tumbling" (e.g. 4179 Toutatis; Mueller et al., 2002), they are in the minority. Comets, however, are often subjected to forces, such as jetting, that can rapidly change their spin state. The approach to the Sun aggravates the problem because much is undoubtedly going on and changing beneath the coma, but it is all but invisible to ground-based observers.

As an example of the difficulty, consider the most famous comet – 1P/Halley. During its 1985/1986 apparition, it was observed by a worldwide network of telescopes as well as a spacecraft, collectively referred to as the International Halley Watch (IHW). In the immediate aftermath of this campaign, Wood (1987) noted that rotation periods of 2.2 and 7.4 days were both possible, with some arguing that one of the two periods was the axial rotation and the other was a nutation or precession – a wobble in the direction of the rotation axis. A more extensive analysis of the photometric data, including all the narrowband photometry, was conducted later by Schleicher et al. (2015) who found the comet to be in a non-principal axis rotation state. Its apparent period varied between 7.2 and 7.6 days, a composite of a long-axis rotation on the order of 7 days long and a ~3.7-day precession period. Its rotational lightcurve flipped back and forth between double- and triple-peak forms over the course of 2 months, and slowly changed in shape and amplitude. They were unable to develop a scenario that included the earlier 2.2-day component period in any way and excluded it. Despite the plethora of observational data and analysis, its exact rotation state is still not completely understood.

6.3 Phase Curves of Planets

In this section, we look at the disk-integrated phase curves for planets represented by two endmember cases: Venus and Mercury. Venus is covered by an optically thick atmosphere, while Mercury has no atmosphere and only minor surface albedo variability. And with the exception of the Moon, both planets have the most extensive telescopic phase coverage in the solar system.

Planetary atmospheres are dominated by molecules and particulates (dust, liquid drops, or ice) separated by distances large compared to the size of the molecule or particulate. This means that the scattering behavior can be

6.3 Phase Curves of Planets

Figure 6.2. Venus transiting the Sun in July 2015. Its atmosphere is evident from the thin rim of light seen against the backdrop of space.
Credit: JAXA/NASA/Hinode.

described by Rayleigh scattering if mostly composed of molecular species, more general Mie scattering if dominated by particulates on the order of the wavelength in size, or geometric optics if many times larger than the wavelength. We generally don't have to worry about self-shadowing or the opposition surge.

Venus is the classic example of a planet with a thick atmosphere, and because it exhibits phase angles from $0°$ to $180°$ when viewed from Earth, has a long history of its study. Its atmosphere was discovered in 1761 by the Russian astronomer **Mikhail Lomonosov** (1711–1765) during a transit of the Sun (Marov, 2005). Lomonosov noted that there was a rim of brightness surrounding Venus even on the portion of the disk that was still off the solar disk, and from this, he correctly surmised the presence of a thick atmosphere (Figure 6.2).

6.3.1 Venus Disk-Integrated Observations and Fit

Figure 6.3 shows the reduced visual-magnitude phase curve for Venus (data extracted from the summary paper of Mallama et al. (2006); See also Figure 2.10.) The data are best described as a curve that is ever-so-slightly non-linear and convex over most phase angles, with a roll-over to a maximum at $0°$. There is also an anomalous brightening at phase angles $>164°$ that has only recently been discovered (Mallama et al., 2006).

As a first order of business, we may wonder whether the phase curve can be fit by a simple model, like the Lambertian or Lommel–Seeliger (L-S)

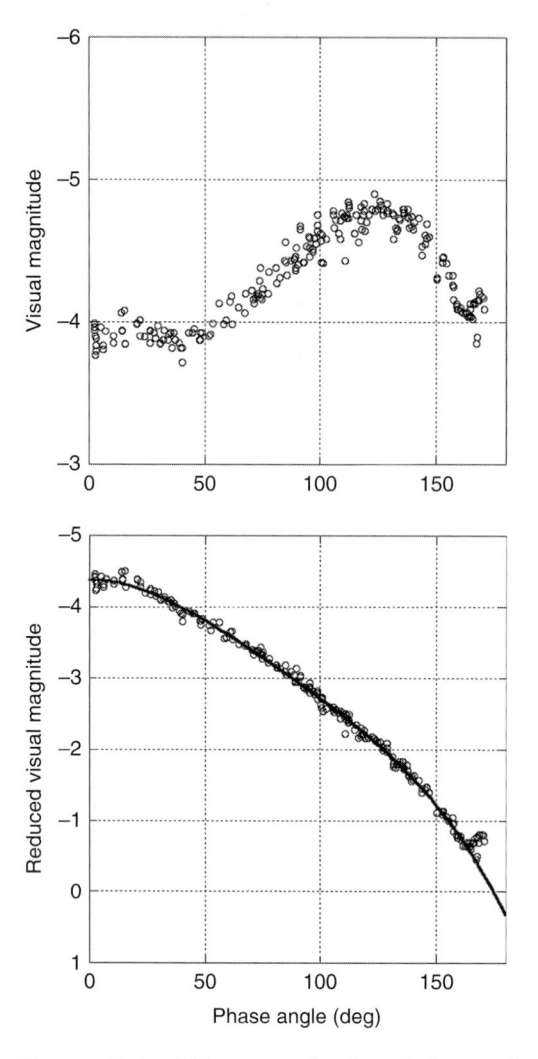

Figure 6.3. The magnitude of Venus as a function of phase angle (top) and reduced magnitude (bottom). The reduced magnitude phase curve is shown with the best fit polynomial (Table 6.2).
Credit: Data from Mallama et al. (2006).

(Chapter 5). Table 6.1 lists Venus's geometric albedo as $p_v = 0.67$. Immediately, we can rule out a simple L-S model because the maximum geometric albedo for an object obeying that scattering law is $p_v = 0.125$ (Table 5.2). The Lambertian model is a possibility because the maximum geometric albedo for that surface is $p_v = 0.67$, meaning that Venus would have to be a perfect

6.3 Phase Curves of Planets

Table 6.1 *Disk-Integrated Properties of the Planets*

Planet	V(0)	p_v	A	D (km)
Mercury	−0.42	0.14	0.068	4,879
Venus	−4.40	0.67	0.90	12,104
Earth	−3.86	0.37	0.306	12,742
Mars	−1.52	0.17	0.250	6,779
Jupiter	−9.40	0.52	0.343	139,822
Saturn	−8.88	0.47	0.342	116,464
Uranus	−7.19	0.51	0.300	50,724
Neptune	−6.87	0.41	0.290	49,244
Pluto	−1.0	0.5–0.7	0.4–0.6	2,374
Moon	+0.21	0.12	0.11	3,474

Lambertian scatterer (Table 5.2). A plot of the Lambertian phase curve with the Venus data show it to work reasonably well, though systematically offset, for phase angles <120°; after that, the Lambertian model grossly underpredicts the actual brightness of Venus.

One might suppose a more sophisticated model such as the disk-integrated Hapke model with a high single-scattering albedo ($w = 0.99$) and no roughness or opposition surge (Chapter 5) would work better, but in fact, it is virtually indistinguishable from the Lambertian model.

For most planets, including Venus, the best model for predicting reduced magnitude at a given phase angle is a simple polynomial, determined empirically, of the form

$$V(1, \alpha) = c_0 + c_1 \alpha + c_2 \alpha^2 + \cdots + c_n \alpha^n \qquad 6.12$$

Using perhaps the most exhaustive data set of Venus disk-integrated observations, Mallama et al. (2006) determined fourth-order polynomial fits for the Venus phase curve in blue (B), visible (V), red (R), and infrared (I) passbands. The fit for the V-band ($\alpha < 164°$) is reproduced in Table 6.2 and shown in Figure 6.3.

What, in broad terms, can be gleaned from this phase curve? The lack of an opposition surge, and in this case the downward turn at opposition, is a distinctive characteristic of photometric surfaces with widely spaced particulates on the order of or larger than the wavelength, i.e. clouds. The other feature of note, peculiar to Venus, is the abrupt brightening at $\alpha > 164°$. Mallama et al. (2006) note that it differs slightly between B and V wavelengths, suggesting a Mie scattering phenomenon (particles of size on the order of the wavelength). Subsequent modeling suggests 2 μm liquid droplets of sulfuric acid (H_2SO_4) best fit all the characteristics of this behavior (Mallama et al., 2006).

Table 6.2 *V-band Phase Functions of the Planets*

Coefficient	Mercury	Venus
c_0	−0.616	−4.38394
c_1	6.617339×10^{-2}	-1.04396×10^{-3}
c_2	-1.867745×10^{-3}	3.68682×10^{-4}
c_3	4.103536×10^{-5}	-2.81374×10^{-6}
c_4	-4.583032×10^{-7}	8.93796×10^{-9}
c_5	2.643953×10^{-9}	
c_6	$-7.012340 \times 10^{-12}$	
c_7	6.592807×10^{-15}	

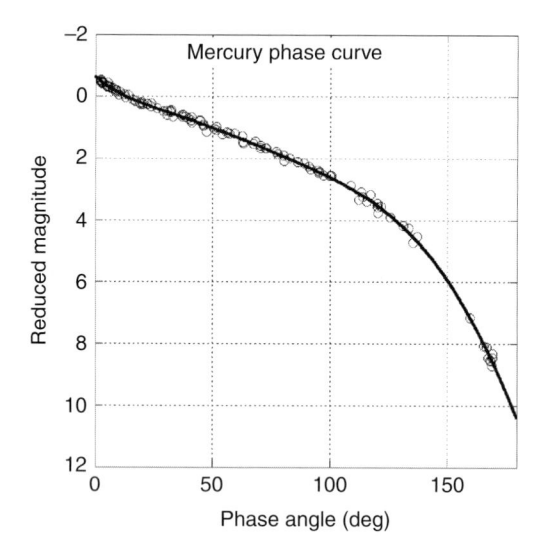

Figure 6.4. Reduced magnitude phase curve of Mercury and best fit polynomial curve fit (Table 6.2).
Credit: Data from Mallama et al. (2002).

6.3.2 Mercury Disk-Integrated Observations and Fit

Mallama et al. (2002) combined spacecraft observations from the Large-Angle Spectrometric Coronograph (LASCO) with extensive ground-based CCD observations to generate a Hermitian phase curve from 2° to 170° (Figure 6.4; see also Figure 4.12). In addition to being much darker than Venus, Mercury's phase curve differs from it in another major respect – it brightens more than 40% as it moves from a phase angle of 10° to 2°, while the illuminated fraction of the disk only increases 1% (Mallama et al., 2001). This rapid change is the opposition surge and is characteristic of a scattering media dominated by closely packed scatterers, as in a regolith. As with Venus, Mallama et al. (2001) find an empirical polynomial equation for the Mercury phase curve in V-band, given in Table 6.2. In Section 6.6, we will examine this phase curve along with others in different wavelengths.

6.3.3 Values for Other Planets

Analyses similar to that shown for Venus and Mercury have taken place for nearly every major object, planet or moon, in the solar system. In Table 6.1, we list the absolute magnitudes, $V(0)$, visual albedos, p_v, Bond albedos, A, and diameters, D, of the more prominent. These were taken from the NASA planetary fact sheets and are standard published values in a variety of references. But while the diameters are accurate to 1% or better, the values for $V(0)$, p_v, and A vary considerably in the literature; one need only insert these values into Equation 6.7 to see that they do not always match exactly. As an example, Hilton (2005) lists estimates of $V(0)$ for Venus found in the literature ranging from –4.11 to –4.91; much of this may be caused by systematic differences in the data sets used, and in the methods used to extrapolate those phase curves to 0°. In other cases, the magnitude may be a function of rotation angle and orbital position. For example, Mars has significant albedo variations on its surface and, depending upon our relative positions, may display more of the northern or southern hemisphere to Earth. It is also subject to dust storms that significantly affect the albedo (Mallama, 2007).

6.4 Phase Curves of Small Bodies

In the previous section, we looked at the phase curves of Venus and Mercury and noted that phase curves for all other major objects were obtained long ago. As we will see in the next chapter, there are occasions when the ground-based phase curves of a planet or moon can be fruitfully paired with high-resolution spacecraft images and surface photometry; but in general, this kind of work on the planets and larger moons is of far less import than in the past. Not so with small bodies. There are millions of main-belt and Trojan asteroids, Centaurs, and icy bodies in the Kuiper belt. Only a few have or ever will be seen in disk-resolved detail. For the foreseeable future, disk-integrated observations are the only source of information about the vast majority of these objects. This is also the case for newly discovered extra-solar planets.

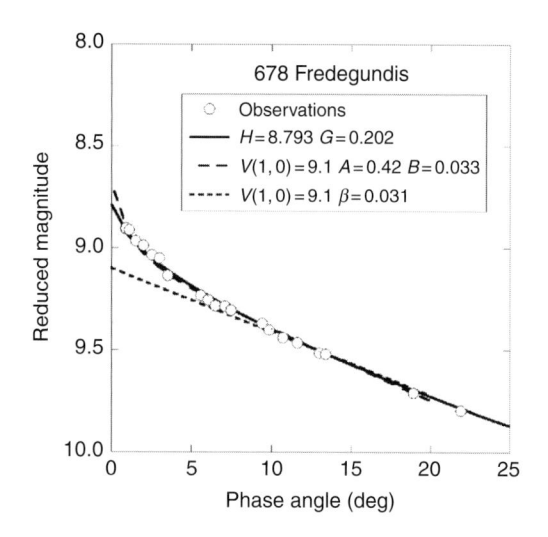

Figure 6.5. Reduced magnitude phase curve of Asteroid 678 Fredegundis and best fits in the Shevschenko, $V(1, 0)/\beta$, and HG system.
Credit: Data courtesy B. Warner.

Given their position relative to the Earth, most main-belt asteroids have phase angles ranging from 0° to a maximum of ~25°. Trojans, Centaurs, and Kuiper belt objects have even narrower phase limits. Only a very few spacecraft-visited asteroids or close-passing near-Earth asteroids have been observed outside of this range. Given that the bulk of the phase curve data lies with main-belt asteroids, we focus on the historical work within those confines. As an example for illustrative purposes, Figure 6.5 shows a phase curve for asteroid 678 Fredegundis.

6.4.1 Lightcurve to Phase Curve

An important question is how to go from a lightcurve to a phase curve. Lightcurves vary in time over periods of hours, while phase curves change over weeks and months. Which part of the lightcurve is chosen to calculate the reduced magnitude? The IAU standard for the HG magnitude system (discussed later) is to use the average over a lightcurve cycle (Bowell et al., 1989). In the language of the Fourier series expansion of the lightcurve (Section 6.2), this would mean using the a_0 component. One reason for choosing the average to define the phase curve is that randomly timed observations of an asteroid will tend toward the mean.

However, Alan Harris (personal communication, October 2015) has argued that if one wishes to use the phase curve data for physical interpretations, it is best to use the data taken at lightcurve maximum. The reasoning is as follows. Most small bodies are not spheres, but irregular objects commonly modeled as ellipsoids. At minimum light, one is looking more or less along the two smaller axes of the ellipsoid and the mean slope of that surface is steeper than for the equivalent sphere, and shadowing will have a more pronounced effect on the phase curve. At maximum light, one is looking along the longest and some combination of the intermediate and shortest axis (depending on the viewing latitude), and the mean slope is closer to that of the equivalent sphere than at any other aspect.

6.4.2 Characterizing the Phase Curve and the $V(1, 0)/\beta$ System

It has been noted, at least since Russell (1916a), that phase plots of most asteroids are approximately linear (in magnitudes) for phase angles $\alpha > 10°$, with brightness decreasing at higher phase angles. At phase angles $\alpha < 10°$, i.e. near opposition, many asteroids show an opposition surge. The cause of this phenomenon has been the subject of many investigations, and two mechanisms for this behavior were previously discussed: shadow hiding (SH) and coherent backscattering (CB; Chapter 4). In the next section, we will briefly look at its effect on asteroid phase curves, but for the moment, we focus on the linear portion of the phase curve.

Perhaps the easiest way to fit the magnitude of an asteroid from phases $\alpha \geq 10°$ is with a straight line. The slope of this decrease is called the **phase coefficient**, β, and, for most asteroids, ranges from approximately 0.02 mag/deg to 0.05 mag/deg[3]. Traditionally, the phase coefficient is measured from the slope of the phase curve between $10° \leq \alpha \leq 20°$, although the actual limits are flexible and vary from author to author. Over the phase angle range 10–20°, Fredegundis drops 0.31 mag, giving it a phase coefficient $\beta = 0.031$ mag/deg.

An asteroid's absolute magnitude is one of its more important physical properties because it incorporates an asteroid's innate ability to reflect light – its visual or geometric albedo – as well as its size, or at least apparent diameter as we've derived in Equation 6.7. Because it is at opposition, there are no visible shadows, so the effects of topography are minimized to the extent possible. Unfortunately, it is rarely possible to directly measure an asteroid at $\alpha = 0$, so one must estimate it by some form of extrapolation.

---

[3] Although the slope is always negative, the phase coefficients are often reported as positive numbers.

Prior to 1985, the standard method of characterizing an asteroid's phase curve and absolute magnitude was to report β and the magnitude, $V(1, 0)$, where the extrapolation of the phase coefficient line crossed $\alpha = 0°$. Because of the opposition surge, this estimate almost always underestimated the true absolute magnitude and created systematic problems for using this value to calculate albedos and diameters. The difference between the linear projection and the actual magnitude at or near $\alpha = 0°$ was reported as the **excess magnitude** at opposition.[4] For 678 Fredegundis (Figure 6.5), $V(1, 0) = 9.1$ and the excess magnitude is ~0.3 mag.

6.4.3 The Opposition Surge

The opposition surge is responsible for the excess magnitude and offers both a problem and opportunity. It is an *opportunity* because if we can understand the mechanism causing the surge, it may be modeled and used to deduce something about the physical properties of the surface. It is a *problem* because the surge makes it difficult to determine the absolute magnitude of an object without physically observing it at zero phase.

The shadow-hiding and coherent backscattering mechanisms each have a number of proposed mathematical forms that can be used when fitting phase curve data. However, these typically use several free parameters and non-unique model solutions to a given data set are often a problem. Perhaps the most robust method for fitting the opposition surge, if the least connected to underlying physical mechanisms, is to use a simple empirical function as described in Chapter 4. A common example is one attributed to Shevchenko (1997) and used by Belskaya and Shevchenko (2000) to investigate the opposition surge behavior of 33 asteroids:

$$V(1, \alpha) = V(1, 0) - \frac{A}{1 + \alpha} + B\alpha \qquad 6.13$$

Here, $V(1, 0)$ is the projected absolute magnitude, and A and B are constants fit to the phase curve observations; A characterizes the amplitude of the surge (the excess magnitude), and B characterizes the linear slope of the curve. In most cases B should be the same or very close to the phase coefficient, β. When applied to the Fredegundis data, we find $V(1, 0) = 9.1$, $A = 0.42$, and $B = 0.0333$ with a fit better than 0.5% everywhere. The value of $V(0) = 9.1 - 0.42 = 8.68$.

---

[4] Because of its importance, we reiterate what was stated in Chapter 2. Throughout this book, we use $m(0)$ and $V(0)$ to mean the **actual** absolute magnitude, although they are estimates. The reader who finds an older absolute magnitude listed as $V(1, 0)$ must be aware that it is a linear extrapolation.

6.4.4 HG Magnitude System

In 1979, the Montreal General Assembly of the International Astronomical Union (IAU), Commission 20 formed a committee to develop a standardized method that would provide better estimates of magnitude at all phase angles, especially at opposition. In 1985, that work was complete and the IAU formally adopted a two-parameter convention for reporting asteroid phase curves called the HG system (Marsden, 1985). A detailed description of this system is given by Bowell et al. (1989); here we outline its derivation before summarizing it.[5]

Start from an assumption of the Lumme–Bowell scattering model (Chapter 5) that the disk-integrated radiant intensity from an asteroid, $I(\alpha)$ can be broken into components, $I_S(\alpha)$ and $I_M(\alpha)$, representing the singly and multiply scattered components, respectively.

$$I(\alpha) = I_S(\alpha) + I_M(\alpha) \qquad 6.14$$

If we normalize this value to that observed at opposition, we generate the integral phase function:

$$\Phi(\alpha) = \frac{I(\alpha)}{I(0)} = \frac{I_S(\alpha) + I_M(\alpha)}{I_S(0) + I_M(0)}$$

$$= \left[\frac{I_S(0)}{I_S(0) + I_M(0)}\right]\left[\frac{I_S(\alpha)}{I_S(0)}\right] + \left[\frac{I_M(0)}{I_S(0) + I_M(0)}\right]\left[\frac{I_M(\alpha)}{I_M(0)}\right]$$

where the last expression is simply a regrouping of terms. The second terms in each product are the integral phase functions for singly and multiply scattered light:

$$\Phi_S(\alpha) = \frac{I_S(\alpha)}{I_S(0)}, \quad \Phi_M(\alpha) = \frac{I_M(\alpha)}{I_M(0)}. \qquad 6.15$$

We then define the parameter G as

$$G = \frac{I_M(0)}{I_S(0) + I_M(0)} \qquad 6.16$$

and

$$(1 - G) = \frac{I_S(0)}{I_S(0) + I_M(0)}$$

---

[5] Some of this history and derivation are given in Bowell, Harris, and Lumme (1989) (hereafter BHL89), a manuscript that was often referenced but unfortunately never published. Some elements of it were incorporated into Bowell et al. (1989), but not all. We thank Alan Harris for providing us with a scanned version.

Now we can rewrite the integral phase function as

$$\Phi(\alpha) = (1 - G)\Phi_S(\alpha) + G\,\Phi_M(\alpha) \qquad 6.17$$

This equation shows that G is actually a partitioning or weighting coefficient that lets us blend two different (integral phase) **basis functions**, Φ_S and Φ_M, and that these are functions of the scattering properties of the regolith. Given the reduced visual magnitude of an asteroid at a particular phase angle, $V(1, \alpha)$, H and G are related by the following (using Eqs. 2.19 and 5.57)

$$V(1, \alpha) = H - 2.5 \log\left((1 - G)\Phi_1(\alpha) + G\Phi_2(\alpha)\right) \qquad 6.18$$

where H is an estimate of the asteroid's **absolute visual magnitude** and G is called the **slope parameter**. Note that we have also changed the subscripts for singly and multiply scattered light from S and M into 1 and 2, respectively, to be consistent with the nomenclature in common use. BHL89 and others have pointed out that H is the *mean* absolute magnitude – the average brightness as it rotates, equivalent to the a_0 term in the Fourier expansion of Section 6.2, and not the lightcurve maximum that some have suggested.

After deriving this functional form, BHL89 used the best existing asteroid phase angle data to develop *empirical* forms for the integral phase functions. The actual form of these final functions is rather serpentine and can be found in Bowell et al. (1989), but accurate approximations adopted by the IAU, valid for $0 \leq \alpha \leq 120°$ and $0 \leq G \leq 1$, are given by

$$\begin{aligned}\Phi_1 &= \exp\left[-3.33 \tan^{0.63}\left(\frac{\alpha}{2}\right)\right] \\ \Phi_2 &= \exp\left[-1.87 \tan^{1.22}\left(\frac{\alpha}{2}\right)\right]\end{aligned} \qquad 6.19$$

These two basis functions are endmembers for all asteroid phase curves in the HG system and their shapes can be seen in Figure 6.6. Both functions are normalized so that $\Phi_1(0) = \Phi_2(0) = 1$.

If $G = 0$, the endmember case where there is no multiple scattering, Φ_1 generates a steeply dropping phase curve with $\beta = 0.043$ mag/deg over the phase angles $10°$–$20°$. If $G = 1$, an unrealistic endmember case where there is no single scattering (or multiple scattering totally dominates), Φ_2 generates a shallow phase curve with $\beta = 0.014$ mag/deg. Bowell et al. (1989) note that asteroids generally fall in the range $0 \leq G \leq 0.5$, with darker classes tending to lower G and brighter classes to higher G as expected from this physical understanding.

Upon discovery, an asteroid magnitude can be converted to a reduced magnitude once its orbit is established. In the absence of any other evidence, its absolute magnitude, H, is estimated and it is *assigned* a slope parameter,

6.4 Phase Curves of Small Bodies

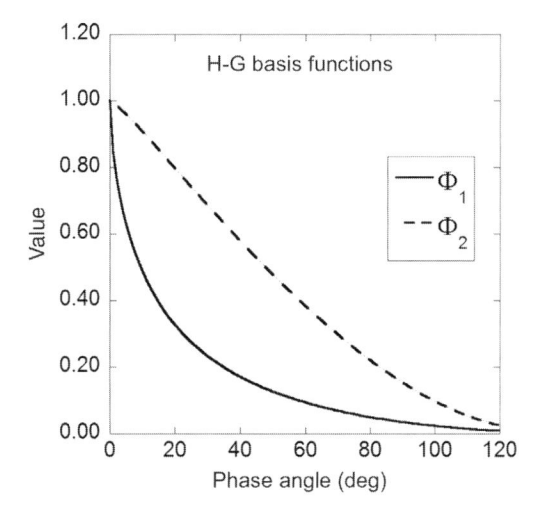

Figure 6.6. A plot illustrating the two endmember cases of the HG system basis functions.
Credit: Michael K. Shepard.

$G = 0.15$. If additional data points on the phase curve are acquired, a least squares fit to the preceding equations can provide a better estimate of H and G. For the observations of 678 Fredegundis, a least squares fit gives a best fit of $H = 8.793 \pm 0.007$ and $G = 0.202 \pm 0.010$. Note our absolute magnitude estimates for Fredegundis range from 8.68 to 9.10 depending upon the method used; the H value seems closer to reality.

One advantage of the HG system over the other methods of reporting phase curve behavior is that G and the phase integral, q, of an object are related by

$$q = 0.290 + 0.684G \qquad 6.20$$

This expression can be found by substituting the integral phase function for the HG system into the equation defining the phase integral (Eq. 5.63). The phase integral, as shown earlier, is the integral of the entire phase function, from 0° to 180°. It, in turn, is related to the geometric albedo, p_v, and Bond albedo, A, by

$$A = p_v q \qquad 6.21$$

Compared with the earlier system of estimating absolute magnitudes (the $V(1, 0)$/ β system), BHL89 point out that the HG system gives estimates that result in geometric albedos about one-third higher (as they should if accounting for the excess magnitude). However, the phase integral is correspondingly lower than that estimate in the older system so that the Bond albedos estimated in both systems are comparable.

6.4.5 HG$_1$G$_2$ Magnitude System

The HG system was adopted because it is a relatively simple two-parameter system that does a good job fitting observed phase curves and is at least heuristically related to the observed scattering properties of asteroid regolith. But there are asteroids that are not well fit by it, especially near opposition. The E-class asteroids have a high albedo and fairly low phase coefficient in general, but a narrow and sharp opposition surge (e.g. Harris et al., 1989). The darker asteroids have higher phase coefficients, but some have little or no opposition surge – their phase curves stay linear (e.g. Shevchenko et al., 2008). This has led to the creation of more complex phase curve models that incorporate three or more parameters (e.g. Shevchenko, 1997; Muinonen et al., 2010; Oszkiewicz et al., 2012). The goal of work like this is to provide better fits to observed phase curves and to provide a set of parameters that might be correlated with asteroid albedo, taxonomic type, or other properties.

In 2012, the IAU General Assembly Division III Commission 15, aka Physical Studies of Comets and Minor Planets, accepted a newer HG$_1$G$_2$ system as a replacement for the HG system (Bockelee-Morvan et al., 2015). This system is similar in form to the HG, but with one additional parameter:

$$V(1,\alpha) = H - 2.5 \log \left(G_1 \Psi_1(\alpha) + G_2 \Psi_2(\alpha) + (1 - G_1 - G_2) \Psi_3(\alpha) \right) \quad 6.22$$

The Ψ basis functions for this system are cubic splines. A spline is a piece-wise mathematical function; that is, it has one form for a given domain of numbers (for example, $x \geq 0$) and a different form for another ($x < 0$), but is smooth at the domain interface. Details of this system and these functions can be found in Muinonen et al. (2010).

6.4.6 Physical Causes of the Phase Curve

One might ask how much of the decrease in brightness with phase angle is expected because less of the illuminated portion of the disk is visible. Recall from Chapter 5 that the relative change in the brightness of a sphere with phase angle is given by its integral phase function, $\Phi(\alpha)$. For a sphere obeying an Area Law (brightness is proportional to visible area), this is

$$\Phi(\alpha) = \frac{1}{2}(1 + \cos \alpha) \quad 6.23$$

Integral phase functions for spheres obeying the Lambertian and Lommel–Seeliger scattering laws are also given in Table 5.1. To directly compare phase curves of a sphere obeying these scattering laws to a typical asteroid,

6.4 Phase Curves of Small Bodies

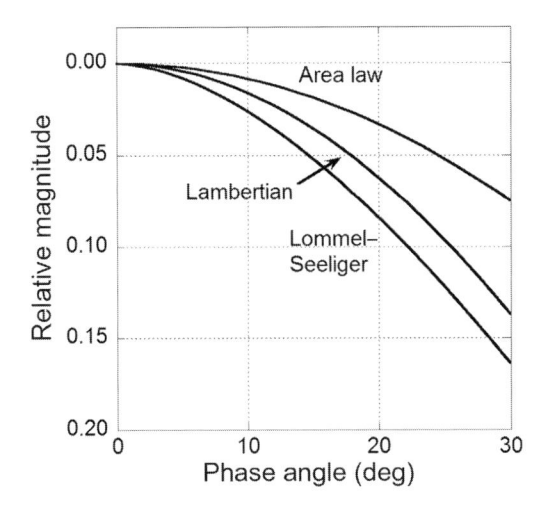

Figure 6.7. Three different scattering functions to illustrate the rate at which their brightness falls off with phase angle. The top curve assumes magnitude drops as a function only of the visible area of the asteroid; the middle assumes a Lambertian scattering function; the bottom assumes a Lommel–Seeliger scattering function. Credit: Michael K. Shepard.

we convert apparent visual magnitudes into reduced visual magnitudes and write (see also Eq. 6.18)

$$V(1, \alpha) = V(0) - 2.5 \log \Phi(\alpha) \qquad 6.24$$

In Figure 6.7, we plot Equation 6.24 for the three scattering functions, arbitrarily setting $V(0)$ to 0. These show at least two things. First, none of these phase curves drop quickly enough to explain most asteroid phase curves (note that Fredegundis drops 0.3 mag in $10°$). Second, these phase curves are all convex (curve downward), while the typical asteroid phase curve is concave, especially around opposition. Of these simple scattering models, the Lommel–Seeliger curve has the largest phase coefficient of $\beta = 0.006$ mag/deg, approximately five times smaller than Fredegundis. Clearly, there are other physical factors needed to explain the observed phase coefficients.

Experiments and modeling suggest two additional factors are necessary to produce a typical asteroid phase curve. First, a more realistic scattering model than the three shown previously is needed. Most particulate surfaces have an opposition surge and stronger fall-off with phase angle than these. Second, we have ignored macroscopic roughness. All objects within the solar system, but especially the smaller objects like asteroids and comets, are extensively roughened and in some cases, quite irregular. This gives rise to significant shadowing not accounted for in the smooth sphere model shown in the plot.

One of the earliest attempts to model these behaviors was by Veverka (1971b). He used a laboratory-measured phase curve of a dark furnace slag that had been found similar to that observed from the lunar surface. Then he adopted a model of a crater-covered surface to determine a shadowing function, and numerically integrated over the surface to estimate the brightness at any given phase angle. He found that typical phase coefficients of 0.03 mag/deg were relatively simple to reproduce given this more realistic scattering function and surface roughening.

Scientists have often noted an apparent correlation between the slope of the phase coefficient or slope parameter and the geometric albedo of the asteroid. Veverka (1971b) traces this discussion back as far back as Bell (1917). The original IAU notice listed guidelines for assigning a preliminary G given an asteroid's classification (Marsden, 1985; Bowell et al., 1989): 0.15 for low albedo types (C, D, G, P, T), 0.25 for moderate albedo types (A, B, M, Q, S), and 0.4 for higher albedo types (E, R, V). These are still reasonable guides, but today, in the absence of any other information, a value of $G = 0.15$ is typically assigned based on the average value for C- and S-classes.

At least some of this correlation might be understood as a difference in the change from darker objects dominated by single scattering behavior to brighter ones dominated by multiple scattering. As we saw, this behavior led heuristically to the HG magnitude-slope system for asteroids, and the statistics bear out the following generalities. Higher albedo asteroids have shallower slopes (higher G) and darker asteroids have steeper slopes (lower G). Although the correlation is not perfect, it has been used to estimate the albedo of an asteroid from its phase coefficient or vice versa. However, both the correlation and its use in this manner have been disputed by Veverka (1971b), who argued that it is not possible to separate albedo and macroscopic roughness effects from a phase curve alone. While this may be true if asteroid surfaces ranged from smooth to quite rough, it is now thought (and spacecraft data back this up) that most asteroids have similarly rough surfaces, making the correlation more reliable (Bowell and Lumme, 1979).

Belskaya and Shevchenko (2000) looked more closely at the relationship between geometric albedo and phase angle behavior, including near opposition, for 33 well-characterized asteroids. Their work seems to confirm an inverse correlation between geometric albedo and phase coefficient with the following best fit function:

$$\beta = 0.013 - 0.024 \log p_v \qquad 6.25$$

where both constants are good to ±0.002. They also claim to find a correlation between the geometric albedo and opposition surge. Here, they measured the amplitude of the opposition effect in two ways: (1) the excess magnitude between the linear phase extrapolation to 0.3° and an observation at that phase angle; and (2) the ratio of light intensity measured at 0.3° to that at 5° phase angle. Both methods of quantifying the surge gave similar results – a low surge amplitude at low albedos ($p_v \sim 0.04$) rising to a peak at moderate albedos ($p_v \sim 0.2$), then dropping to low amplitudes at high albedo ($p_v \sim 0.5$). They suggest this behavior is best explained by the combined effects of shadow hiding, dominant in darker objects, with coherent backscattering, dominant in brighter objects.

6.5 Polarization Phase Curves

The vast majority of disk-integrated telescopic observations are in filtered light with no polarization discrimination. However, the sense of polarization in scattered light is a powerful tool for extracting information about an asteroid or comet surface, or cometary coma and tail. It is more difficult to gather this information, but often worth the effort.

6.5.1 Conventions

As described in Chapter 2, sunlight is unpolarized. Upon scattering or reflection, however, it can be partially polarized, depending upon the medium and geometry. In planetary astronomy, radar astronomers measure the circularly polarized components, while optical astronomers tend to measure the linearly polarized components. Here, then, we focus only on the linearly polarized components.

The scattering plane is defined to be that containing the incident and scattered ray of light. In Chapter 4, we initially introduced the polarization terms horizontally and vertically polarized light, before noting that most optical astronomers use perpendicular and parallel, respectively, in place of these, where the term refers to the orientation of the electric field vector with respect to the scattering plane (e.g. perpendicular polarization means the electric field vector is perpendicular to the scattering plane). With yet another terminology change to note, when scientists describe the polarization of light from an asteroid or planet, the *convention* is to refer to polarization perpendicular to the scattering plane, as a **positive polarization**, while that parallel to the scattering plane is referred to as **negative polarization** (Geake et al., 1984). In the Stokes vector notation, it is the Q, or S_1 term that indicates this: for horizontal or positive polarization, $Q = 1$, while for vertical or negative polarization, $Q = -1$.

We measure the intensities[6] of light polarized in these directions as $I_\perp$ and $I_\parallel$ and define the **degree of linear polarization** (hereafter referred to simply as the **degree of polarization** – linear being implied) as

$$P = \frac{I_\perp - I_\parallel}{I_\perp + I_\parallel} \qquad 6.26$$

For most astronomical objects, the degree of polarization is reported as percent or as permil (parts per thousand).

6.5.2 Observations and Calibration

6.5.2.1 Laboratory

When experimenters began to measure the degree of polarization from particulate surfaces thought to be analogous to the surfaces of the planets, they found a fairly standard behavior, illustrated in Figure 6.8.

At opposition, there is no polarization (there shouldn't be since the planes are undefined here). As the phase angle increases, the degree of polarization becomes negative, reaches a minimum, called P_{min} at phase angle α_{min}, and begins to steadily increase, crossing zero polarization at a phase angle referred to as the inversion angle α_{inv} (α_0 and V_0 are also used). This portion of the phase curve makes up the so-called **negative branch**, and α_{inv} is its width. After crossing the zero-polarization axis, the polarization curve is approximately linear with slope h, and remains approximately linear for tens of degrees before peaking at P_{max} and descending again. This is the **positive branch** of the polarization curve. For observations of most objects in the solar system, phase angles are restricted such that P_{max} is never reached, and investigators tend to focus on the negative branch. The primary quantities of interest are P_{min}, α_{inv}, and h.

A number of experimentalists, conducting extensive laboratory work on lunar samples, meteorites, and other particulate materials, have found several

---

[6] Although the sense of polarization may be referred to as positive or negative, *intensities* are always positive in magnitude.

6.5 Polarization Phase Curves

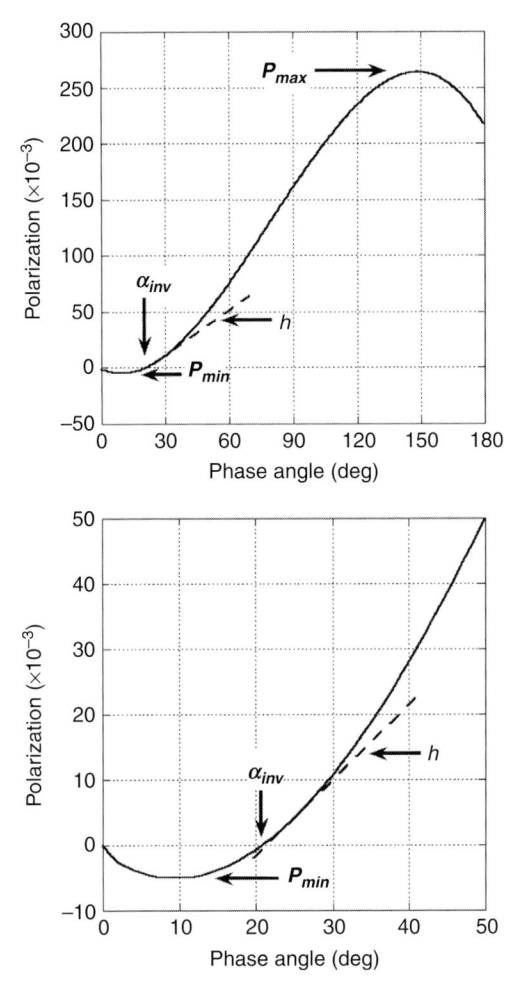

Figure 6.8. A representative polarization phase curve with the major parameters of interest noted. The top figure shows the entire curve, the bottom figure is an enlargement and covers the range of phase angles available for most solar system objects.
Credit: Michael K. Shepard.

empirical relationships between these polarization parameters and the characteristics of the regolith, including grain size and the geometric albedo. The findings can be summed up in a few statements (Dollfus and Geake, 1977; Geake et al., 1984; Geake and Dollfus, 1986) (Figure 6.9).

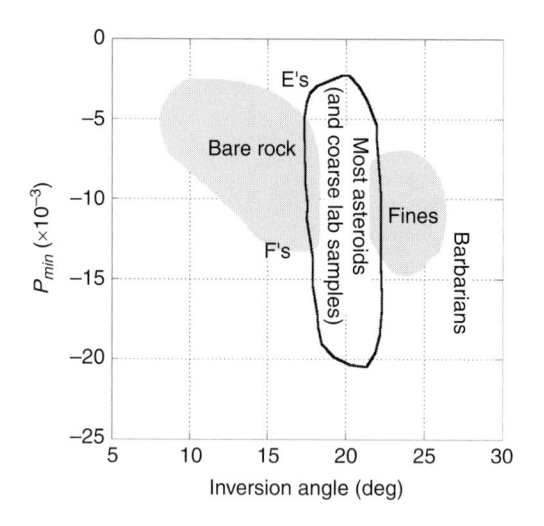

Figure 6.9. Range of polarization minima and inversion angle for a variety of laboratory samples and solar system objects. All minima (P_{min}) are given in terms of parts per mil (10^{-3}) and inversion angles (α_{inv}) in degrees. Regions show behavior of laboratory samples of bare rock and fines. Typical asteroids fall in the indicated region; this also corresponds to laboratory measurements of coarse samples (sizes 30–300 μm).
Credit: After Geake and Dolfus (1986) and Belskaya et al. (2015).

1. The inversion angle, α_{inv}, appears to be an indicator of the coarseness of the regolith; solid fragments have the lowest inversion angles, and finely powdered dust the highest.
2. The slope at the inversion angle ($\Delta P/deg$), h, is a function of the albedo of the sample; high albedos have the lowest slope and low albedos have the highest slope.
3. The depth of the negative branch, P_{min}, is also a function of albedo; lower albedos have more negative values of P_{min}.
4. A less developed finding is that P_{max} is inversely proportional to albedo, and also appears to be a function of the grain size.

6.5.2.2 Astronomical Observations

Based on the success of the early laboratory work, there has been great interest in the use of polarimetry to characterize an asteroid's geometric albedo, and a significant amount of work done to calibrate useful empirical relationships (e.g. Dollfus et al., 1989). To characterize the polarimetric phase curve for typical asteroid, the following relationship has been found to work well (Muinonen et al., 2009; Belskaya et al., 2015; Cellino et al., 2015, 2016):

$$P = A\left(e^{-\left[\frac{\alpha}{B}\right]} - 1\right) + C\alpha \qquad 6.27$$

where A, B, and C are constants determined by a best fit to observations. The relationship between geometric albedo, p_v, and the slope, h, or P_{min} is found to be well described by equations of the form (Masiero et al., 2012; Cellino et al., 2015, 2016)

$$\log p_v = A \log X + B \qquad 6.28$$

where X is either h or P_{min}, and A and B are constants determined by fits to asteroids with well-calibrated geometric albedos and polarization phase curves. Based on fits to approximately 100 such asteroids, Masiero et al. (2012) published the following empirical relationships, where the constants are good to about one significant figure:

$$\log p_v = -1.2 \log h - 1.9 \qquad 6.29$$

and

$$\log p_v = -1.6 \log |P_{min}| - 0.9$$

where P_{min} is given in percent.

For *most* asteroids, the value of P_{min} ranges from -0.4% to -2.3%, the inversion angle α_{inv} typically falls between 18° and 23°, and the slope of the curve at the inversion angle h ranges from 0.05%/degree to 0.3%/degree.

6.5.2.3 Oddities

Not all asteroids obey the rules just described. The brighter asteroids, typically E-class such as 44 Nysa, have extremely shallow slopes and small P_{min}, making it difficult to have confidence in extracted geometric albedos. Several also exhibit a separate narrow and unexpected P_{min} spike at tiny phase angles, possibly related to the coherent backscattering opposition surge (Rosenbush et al., 2002, 2009).

Some darker asteroids, typically of the F-class (704 Interamnia being the classic example), have inversion angles ~16°, much smaller than expected (Cellino et al., 2015, 2016), and a number of asteroids have inversion angles approaching 30°, much higher than normal. The first asteroid discovered with this property was 234 Barbara, so objects displaying this anomalous behavior are referred to as **Barbarians** (Cellino et al., 2006, 2015, 2016). One suggested cause of this anomalous behavior is the presence of spinel, a mineral with a high index of refraction common in calcium–aluminum inclusions that might be found on the surface of these asteroids (Belskaya et al., 2015).

6.5.2.4 Comets

Comet nuclei exhibit polarization phase curves similar to those of asteroids. The dust in the coma, however, provides other opportunities and challenges. Unfortunately, the observed scattering behavior of the coma changes with phase angle because of its intrinsic scattering properties, but also because cometary activity, and these two factors are not always easy to separate. An overview of the problems and methods to address this problem are found in Kolokolova et al. (2004).

Because of its extended nature, there may be some opportunities for investigating cometary comae with polarization. One opportunity, not well explored, is the behavior of P_{max}. As we noted, P_{max} occurs at phase angles ~90° or more, and is not applicable to most asteroid observations. However, numerous comets have been observed at these high phase angles with polarized light, and Kolokolova notes that comets separate naturally into groups with higher and lower values of P_{max}. Some early laboratory work by Geake and Dollfus (1986) have shown that P_{max} is a function of albedo and grain size, but additional work is necessary for this to be of use.

6.6 Case Study: Mercury

Mercury is a difficult planet to study. It is small – not much larger than our Moon – and it is quite close to the Sun, at least from the perspective of the Earth. It is never more than 28° away from the Sun, so it is always in twilight or daylight, and this makes it a technical challenge for telescopic observations. Despite this, its photometric behavior was extensively studied by Gustav Muller (Potsdam Observatory, Chapter 2) from phase angles of 50°–120° (Muller, 1893; Warell and Bergfors, 2008). The next major investigation was conducted by the French astronomer **Andre-Louis Danjon** (1890–1967) while Director at the Paris Observatory (Danjon, 1949, 1953); he extended the phase angle coverage from 3° to 123°. The most recent work, discussed here, is the extensive V-band phase curve of Mallama et al. (2002) and the five-filter UBVRI phase curves of Warell and Bergfors (2008).

6.6.1 V-band Phase Curve

Mallama et al. (2002) combined the data of two instruments: the Large Angle Spectrographic Coronagraph (LASCO) on the Solar and Heliospheric

Observatory Spacecraft (SOHO) at the Earth's L1 Lagrangian point, and a ground-based 0.2 m f/10 Schmidt–Cassegrain stopped down to 0.045 m aperture. Because SOHO was designed to look at the Sun, it was well suited to make observations of Mercury when at inferior conjunction ($\alpha \sim 180°$) and superior conjunction ($\alpha \sim 0°$). The ground-based telescope filled in the remainder of the phase curve.

The LASCO instrument did not have a V-filter, but did have filters in the blue (B, ~400–550 nm) and orange (O, ~520–640 nm). These observations were transformed to V-band magnitudes in the following manner. Forty-eight bright stars, with a variety of colors (as evident from their B–V indices) were identified in the LASCO image archives, each with well-known V-band magnitudes ranging from 1.2 to 5. Each was measured in the LASCO B- and O-filter and their B–O color indices tabulated. Because the V-band is between B and O, it is reasonable to use linear regression to interpolate between these filters to estimate the V-band magnitude. Instrumental B- and O-magnitudes ($-2.5 \log(counts)$) were measured from the images, and color indices were plotted and fit with a linear equation of the form

$$V - O = c_1(B - O) + c_2 \qquad 6.30$$

where the V-magnitudes are taken from the Yale Bright Star Catalog. Now, given the LASCO B- and O-magnitudes of Mercury (and the best-fit constants c_1 and c_2), one can find its V-magnitude and, from that, reduced V-magnitudes.

The ground-based telescope was fitted with a Johnson V-filter and used a CCD camera with a Texas Instrument TC-211 chip (192 (horizontal) × 165 (vertical) pixels, each 13.75 μm (horizontal) × 16 μm (vertical) in size). There were two complications in measuring V-magnitudes with this system. The first is that Mercury is not a point source, but a disk, and the second is that many of the measurements were made with significant daylight (seeing is better when Mercury is higher in the sky) and subtraction of the background sky becomes critical. To solve this issue, the authors used a modified aperture method. Here the counts of Mercury and comparison star were measured as a function of the aperture radius. As the radius increased, the ratio of counts approached an asymptote that was a function of only the ratio of the relative brightness of the star and Mercury.

The phase curve of Mercury was fit with a seventh-order polynomial (Table 6.2 and Figure 6.4) and, separately, a Hapke model to specifically compare the opposition surge and photometric roughness of Mercury with those of the Moon. In general, those parameter solutions were comparable to those of the Moon.

6.6.2 UBVRI Phase Curves

Warell and Bergfors (2008) conducted a comprehensive ground-based campaign designed to measure the phase curve of Mercury in the five Johnson-Cousins filters, largely in preparation for the Messenger and Bepi Columbo missions to Mercury. They used a 0.9 m $f/10$ Cassegrain telescope stopped down to 0.3 m with special masks and extension tubes to minimize stray light and observations at small phase angles. In all, they report between 12 and 19 observations in each filter, with phase angles ranging from 22° to 138°, and up to 152° in three filters. These observations were fit with a six-parameter Hapke model (w, h, B_0, two-term H-G function (b and c), roughness parameter). Because their data only went down to $\alpha = 22°$, they could not constrain the width of the opposition surge, h, or the H-G backscattering parameter (c), so they used prior solutions (Warell, 2004) in the V-band that incorporated the wider Mallama et al. (2002) phase curve, and fixed both h and c. The other parameters were allowed to float.

In addition to providing data and model fits that would later be used by the MESSENGER mission for photometric calibration, the authors used this study to look for evidence of **phase reddening** in Mercury's scattering behavior. Phase reddening is the observation that an object becomes brighter in longer

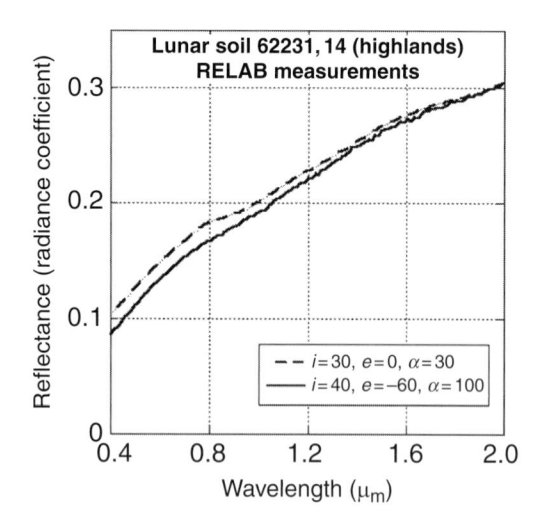

Figure 6.10. Spectra of a lunar sample measured by RELAB illustrating phase reddening. Note that at the higher phase angle, the sample is darker in the shorter wavelengths while retaining the same radiance coefficient at longer wavelengths. Credit: Michael K. Shepard and RELAB Spectral Database. Data acquired by C.M. Pieters. File LS_CMP_30_62231.

wavelengths (i.e. it *reddens*) with increasing phase angle. It has been observed on the Moon (e.g. McCord, 1969), asteroids (e.g. Taylor et al., 1971), and in the laboratory (e.g. Schroder et al., 2014). Figure 6.10 shows phase reddening in the spectrum of a sample of Apollo regolith taken with the RELAB spectrometer (Chapter 7) at two different phase angles. There is a related phenomenon, more rarely observed, referred to as **phase bluing** that has been seen on Mars and some asteroids (Guinness, 1981; Rosenbush et al., 2009) and in the laboratory (Schroder et al., 2014).

Warell and Bergfors measured the amount of phase reddening in three ways: (1) the change in the best fit phase coefficient, $\beta(\alpha < 90°)$ of the raw phase curve as a function of wavelength; (2) the phase coefficient of the Hapke fit to the phase curve as a function of wavelength; and (3) average changes in four color indices (*U–B, B–V, V–R, R–I*) with phase angle. The color indices were normalized to be comparable to methods (1) and (2) using

$$CI_N = \frac{m_{short\ \lambda} - m_{long\ \lambda}}{\lambda_{short} - \lambda_{long}} \qquad 6.31$$

All three methods give a measure of phase reddening in units of mag/deg/μm. Although each method showed evidence of phase reddening, the uncertainties were comparable in scale to effect itself.

7
Planetary Disk-Resolved Photometry

Until the mid-twentieth century, the vast majority of planetary photometry was ground-based and disk integrated, with the notable exception of lunar photometry. Spacecraft missions in the 1970s changed this: Viking Orbiters at Mars and Voyager flybys to the outer planets ushered in a new era of high-resolution planetary imaging at a variety of incidence, emission, and phase angles.

In this chapter, we survey disk-resolved planetary photometry. We begin by looking at laboratory methods and facilities before examining the unique issues relevant to spacecraft observations. Finally, we look at a few published studies to illustrate the current state of the art.

7.1 Laboratory Methods and Tests

The bulk of this book has been spent on methods for obtaining and analyzing photometric data from ground-based telescopes or space-based cameras. But in order to determine the composition and physical state of the surface, these kinds of investigations need laboratory studies to anchor physical interpretations. In this section, we briefly introduce the wonderful world of laboratory photometry specific to planetary studies.

7.1.1 Standards

Every laboratory measurement, like every telescopic or spacecraft datum, is most useful when cast in the form of a comparison to a standard, and the most common standard used is a perfect Lambertian surface. As described in Chapters 1 and 5, this surface meets three conditions.

1. It scatters 100% of the incident light.
2. For a constant emission angle, the reflected radiance falls off as the cosine of the incidence angle.
3. For a constant incidence angle, the reflected radiance falls off as the cosine of the emission angle.

In a laboratory setting, it is typical to measure the reflected light from a test surface, using whatever units the detector reports, and compare that to the reflected light (using the same detector units) from a Lambertian standard. Just as there are no perfect circles in nature, there are also no perfect Lambertian standards. However, there are some materials that come close.

One of the earliest standards used was a surface painted with white barium sulfate paint (Ba_2SO_4), or a glass surface coated with magnesium oxide smoke (e.g. Hapke and van Horn, 1963). In the visible and near-infrared region, their spectrum is approximately flat and reflects ~98% of the incident light. The most common modern standard is a sintered (heat-packed into a solid mass) polytetrafluorethylene (PTFE or Teflon™) product. Trade names for this substance include Spectralon®, Fluorilon-99W™, and Zenith Polymer®. These substances reflect ~99% of the incident light over the entire visible and near-infrared spectrum. Over a fairly wide range of incidence, emission, and phase angles, these standard materials are good analogs for Lambertian scattering. However, at higher incidence and emission angles, there are deviations from true Lambertian scattering behavior. There have been a number of studies where these deviations are quantified (Jackson et al., 1992; Bhandari et al., 2011; Figure 7.1).

There are also a variety of products that purport to be gray-scale Lambertian standards. These are produced by mixing small amounts of carbon to the PTFE and are used where one might wish to have one or more spectrally flat Lambertian scattering surfaces with known (but less than 100%) reflectances (Shaw et al., 2016).

7.1.2 Goniometers

In all disk-resolved photometry, reflectance (via some standard) is measured at a known illumination and viewing geometry, and the resulting parameter is a bidirectional reflectance. Technically, the term bidirectional implies collimated light from the source and collimated light into the detector. More properly, what are actually measured are **biconical reflectances** since both the source and detector emit and measure light rays with some measure of divergence. Unless the divergence is significant, though, most lab and spacecraft measurements are referred to as bidirectional.

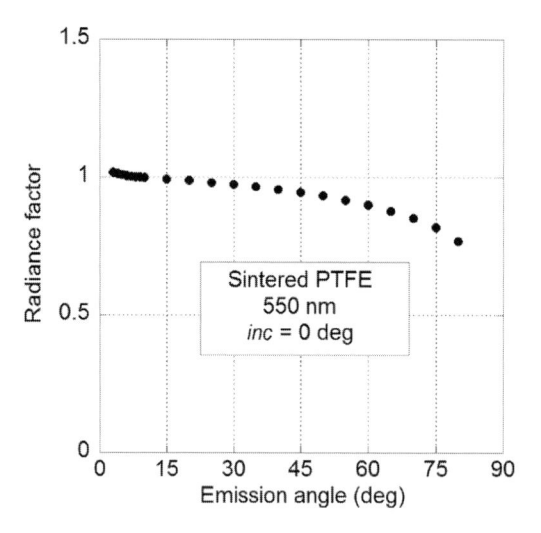

Figure 7.1. Bidirectional reflectance of a laboratory sample of sintered PTFE, a typical Lambertian standard. Incidence angle is 0° and all values are relative to the reflectance at $e = 10°$. Note the fall-off in radiance factor with emission angle. A perfect Lambertian standard would be constant.
Credit: Michael K.Shepard.

7.1.2.1 Mechanics

A **goniometer** is an instrument capable of measuring angles. In photometry, it refers to an instrument capable of illuminating a sample from one direction and measuring the reflected light in another direction. To our knowledge, all photometric goniometers are custom experimental devices.

The most common arrangement for a photometric goniometer places a source and detector at the ends of pivot arms (or some mechanical equivalent, e.g. overhead rails) with a common center, and the sample is placed at the center. In this way, as the source arm pivots, the beam of collimated light stays incident on the sample; likewise, as the detector arm pivots, the detector continues to look at the sample. Many photometric goniometers limit the motion of the source and detector to a vertical plane – the **scattering plane**. A few allow the source and detector to rotate relative to each other in azimuth so that scattering out of this plane can also be measured.

Mechanically, there are some restrictions. The source and detector cannot occupy the same space at the same time, so measurements exactly at opposition are generally not possible. Because the opposition surge is a phenomenon of great interest in planetary photometry, the goniometer is usually designed to go to as small a phase angle as possible. One way is to make one pivot arm longer than the other and arrange to make the mechanism (source or detector) on the shorter arm as small as possible so it does not impede the view or light from the mechanism on the longer arm. In this way, measurements down to a few hundredths of a degree in phase angle can sometimes be obtained. Similarly, measurements at high incidence and emission angles are often desired to test the limits of scattering models. These too require creative engineering to prevent the detector from looking directly at the source in some configurations.

Noise from background lights can be problematic, even if efforts are taken to black out the room. A common solution to this problem is to place a mechanical **chopper** – a rotating mechanical shutter – over the source so that it illuminates the sample with a known frequency, e.g. 30 Hz. The electronics that control the chopper are **synchronized** or **locked in** to the detector electronics so that they measure only the reflected light with the same frequency and phase as that of the source. In this way, background light is excluded and the instrument can safely (and accurately) work in low to moderate light conditions.

Commonly, the laboratory goniometer will ratio all sample measurements to the reflected light from a Lambertian standard. In some devices, this standard measurement is made at the beginning of the run at one geometry (e.g. $i = 0°$, $e = 10°$) and compared with all subsequent sample measurements; in this case, it is important that the light source be very stable. In other devices, the standard measurement is made at the same geometry as the sample, either immediately before or after a sample measurement, in which case light source stability is less critical, but the non-Lambertian properties of the standard must be taken into account.

7.1.2.2 Example Facilities and their Purpose

Early photometric goniometers were usually little more than a source and detector arranged on a horizontal optical bench with some mechanism for changing angular relationships. More sophisticated instruments were built beginning in the 1960s. An early one used in planetary sciences was designed and built by Bruce Hapke and Hugh Van Horn (then at Cornell) to investigate the scattering properties of lunar analog samples (Hapke and Van Horn, 1963). Hapke later improved on this and built a dedicated instrument at the University of Pittsburgh. That mechanism put the source and detector on a semicircular rail above the source (there were no pivot arms), so all measurements were restricted to the scattering plane. It provided a way to produce and measure linearly polarized light (making it a polarimeter as well) and was used to test the early Hapke scattering model (Hapke and Wells, 1981).

Several goniometers have been designed and built specifically to study the opposition surge. Shkuratov et al. (1999, 2002) designed and built a device that used multiple broadband filters and measured the linearly polarized components of reflected light at phase angles down to 0.2°. Nelson et al. (1998, 2000) designed and built the JPL long- and short-arm goniometers. These used linearly and circularly polarized laser sources in two colors, and the long-arm goniometer could reach phase angles down to 0.05°. Kaasalainen (2003) designed and built a device to measure samples at phase angles down to 0.0° using a clever beam-splitter arrangement and Naranen et al. (2004) used this device to measure scattering behavior of materials in a microgravity environment (allowing highly porous samples) by deploying it on a parabolic flight campaign!

The Keck/NASA Reflectance Experiment Laboratory (RELAB)[1] was designed to measure high-resolution spectra of samples from the ultraviolet to the mid-infrared at phase angles ranging from 10° to 140°. The emphasis is on high-resolution spectra, so each measurement at a given geometry may take up to 2 h. For this reason, few samples are measured at more than a single standard geometry (e.g. $i = 30°$, $e = 0°$).

Pinet et al. (2001) built a goniometer that used 19 narrow band filters and an $1{,}152 \times 1{,}242$ CCD array as the detector; in many ways, it mimicked a spacecraft camera. Phase angles ranged from ~20° to 120°. An advantage of this device is that photometry of homogeneous samples or complex scenes, such as craters in layered terrain, can be investigated in some detail (Cord et al., 2003, 2005).

Shepard (2001) designed and built a device to rapidly measure the full bidirectional reflectance properties of samples in multiple broad-band filters. The device used three independent axes of motion (incidence, emission, and azimuth) with computer-controlled stepper motors to measure sample reflectances at phase angles ranging from 3° to 140°. A similar instrument (referred to as PHIRE-1) was built by Gunderson et al. (2006) and employed to study several planetary analog materials. The same group built a more sophisticated goniometer (PHIRE-2) that operates in a sub-zero temperature environment chamber to study the scattering properties of ices (Pommerol et al., 2011).

7.2 Spacecraft Observations

In Chapter 3, we looked at the mechanics of ground-based observation and data reduction. There are a number of important differences between the relatively well-developed telescopic methodology and reduction of modern spacecraft observations.

---

[1] www.planetary.brown.edu/relabdocs/relab.htm

7.2.1 Physical Differences

The space environment is both a blessing and a curse for the spacecraft sensor. A good discussion of the issues can be found in Howell (2006) and we summarize many of those points here.

7.2.1.1 Resolution or Instantaneous Field of View

Earth-based sensors constantly struggle against the atmosphere. Even the best observatories have seeing limitations of ~0.4 arcseconds/pixel, comparable to the theoretical resolution of a 0.3 m telescope. In space, seeing is not a factor and the point-spread function (*PSF*, Section 3.2.2) of stars and planetary surfaces is diffraction limited. To optimize telescope and camera performance requires one to use smaller instantaneous field of view (*IFOV*) in spacecraft cameras than typically needed on Earth.

For a given focal length, *IFOV* can be reduced by using smaller pixels. In principle, this sounds like an excellent solution, but there are tradeoffs to consider because smaller pixels generally mean smaller full well capacities, smaller dynamic ranges, and smaller signal-to-noise ratios (SNRs).

Alternatively, for a given pixel size, *IFOV* can be reduced by increasing focal length. The tradeoff here is a higher magnification and narrower field of view, neither of which may be desirable depending on the application.

7.2.1.2 Mechanics

Most ground-based cameras use full-frame CCDs and mechanical shutters and filter wheels. If there is a problem, a technician takes the camera out and fixes or replaces parts as needed. This luxury does not exist in spacecraft sensors, with the notable exception of the Hubble Space Telescope (and there only at great cost!).

To avoid using a mechanical shutter, many spacecraft CCDs use a frame-transfer CCD. The cost is higher in hardware, but safer. There are also numerous examples of spacecraft CCDs that use the pushbroom architecture. The tradeoff here is increased post-processing and a need for excellent spacecraft pointing and position information to minimize distortion in the final images.

There are also examples of pushframe architecture, like the LRO Wide-Angle Camera. Recall (Chapter 3) that it uses a square $1,024 \times 1,024$ array, but divides the space on the chip between two telescopes and covers subset frames with color interference filters to provide a multi-color image with no moving parts.

7.2.1.3 Space Environment

Cameras on Earth rarely experience temperature swings larger than $10°C$ in background ambient temperature while operating, and are cooled to a stable operating temperature using a circulating coolant (e.g. liquid nitrogen), thermoelectric coolers, or some combination. Spacecraft cameras have operated in orbit around Mercury and as far away as Pluto – an enormous range of thermal conditions. More importantly, there is no buffering atmosphere to keep temperatures relatively stable, and on a spacecraft, the orientation with respect to the Sun is often a matter of electronic life and death. Camera and sensor temperatures must be monitored and stabilized, for years, to have any hope of obtaining quantitative data.

In the vacuum of space, composite materials often deteriorate – sometimes rapidly. Many components slowly outgas volatile compounds such as lubricants. Once volatilized, these substances can inadvertently coat surfaces and compromise the optics.

Without a protective atmosphere or magnetic field, spacecraft are subjected to an onslaught of high-energy particles from the Sun and cosmic background sources. These will often show up in images as bright or hot pixels – extra noise. Some of these can be identified by software (they are typically several standard deviations of the noise level) and removed. But over time, the steady exposure to this higher radiation takes its toll and degrades the performance of the CCD. And without access to laboratory standards for calibration, accounting for that degradation is not easy.

7.2.1.4 Data Compression

Spacecraft are usually power and bandwidth limited. Multiple instruments often compete for the same transmission time, and the great distance from Earth and limited power means transmission data rates are often slow. Most instruments have some method for compressing information to reduce the needed bandwidth. Some methods are **lossless**, i.e. all data can be eventually reconstituted in its original form, but others are **lossy** and accept some loss of information. For imaging information that is not critical for photometric analysis, images are often compressed using a lossy algorithm.

As an example, assume the onboard camera digitizes pixels using 12 bits, giving pixel digital numbers (DNs) of 0–4,095. A simple linear (lossy) compression scheme is to divide the DN by 16 (2^4) onboard the spacecraft, and transmit the result as an 8-bit number with DNs of 0–255. More commonly, several non-linear compression schemes are available; these are often designed to provide better resolution in the mid-range (e.g. divide by 8 for pixels in the middle range) and poorer resolution at the high and low ends.

7.2.1.5 Processing

For US and some international missions, the US Geological Survey (USGS) Astrogeology Branch in Flagstaff, Arizona, has developed a software package

(Integrated Software for Imagers and Spectrometers, or ISIS) that helps users manage and work with spacecraft imaging data. The data are initially formatted by the mission for inclusion into the NASA Planetary Data System (PDS) and are categorized into differing levels, depending on the amount of processing that has taken place.

Level 0 processing is the lowest level and typically involves the formation of a file and extraction of necessary instrument data from the SPICE kernels (see the following section) into headers attached to the imaging data file. **Level 1** processing removes systematic or other known artifacts such as inhomogeneities in pixel response or shading (flat-fielding), noise (dark and bias frames), and smear. The output at this level should be radiometrically calibrated. **Level 2** processing geometrically corrects the image and projects it into standard map coordinates. Some missions create higher level products that may include photometric corrections, compilation of individual images into mosaics, or others.

7.2.2 SPICE

Consider the data pipeline for a moment. An imaging camera on a spacecraft orbiting a distant planet takes an image. Those data are transmitted to Earth and received by the mission scientists. They understand how to take the raw numbers, correct them for dark frames, smear, etc., and calculate a radiance value. But this is only the beginning. Where was this image acquired? How does it relate to images taken yesterday, or that will be taken tomorrow? What was the illumination and viewing geometry at the time of the image? What was the ground resolution? All of these questions require ancillary data that are not part of the image.

7.2.2.1 Origins

Up until around 1990, mission engineers would provide as much of the ancillary information to the scientists as possible, but it was an ad hoc process that wasn't very user friendly, and an outside investigator would have had a difficult time using any data that hadn't already been heavily processed by someone directly involved. Beginning in 1984, a new system, called SPICE, was developed to handle this problem. **SPICE** is an acronym for **Spacecraft Planet Instrument Camera-matrix Events**, and it is the responsibility of the Navigation and Ancillary Information Facility (NAIF) at NASA's Jet Propulsion Laboratory. It is incorporated by default in NASA missions since the 1990s. It is also utilized by many European Space Agency (ESA) missions (e.g. Mars Express, Rosetta) and on a few missions of the Japanese (JAXA) and Indian (ISRO) space agencies.

What is SPICE? It is a collection of software files, called **kernels**, which contain information on the reference frames for the spacecraft, its instruments, and the Earth and solar system (e.g. J2000), the positions and orientations of the spacecraft and instruments within those frames during data acquisition, models for the size and shape of the object being studied, events on the spacecraft or ground during acquisition that may affect results, pointing directions and field of view, and time conversions (e.g. mission elapsed time to *UTC* etc.). In other words, it contains any information that might be needed to put the image into its proper context in time and space so that it may be compared with every other image. (We have focused on images, but SPICE is used for most instrument packages, not just the imaging.) These data come from a variety of sources: the spacecraft and instrument builders, mission control center, spacecraft status data, and organizations like the International Astronomical Union for standards of planetary models (e.g. size, shape, definitions of longitude and latitude, etc.).

7.2.2.2 Data File Types

An excellent overview and set of tutorials on SPICE is available from the NAIF (naif.jpl.nasa.gov). Here we summarize the data file types one might encounter as described by those tutorials.

> Spacecraft **SPK** files contain the trajectory of the spacecraft, ephemerides of the targets, and any other gross-scale positioning information that might be needed.
>
> Planet **PcK** files contain the target shapes, orientations, sizes, and potentially other parameters such as a gravity or atmospheric model.
>
> Instrument **IK** files contain the information specific to the instrument involved such as field of view, orientation, and internal timing.
>
> Camera-matrix **CK** files contain spacecraft attitude and orientation information, relative to the target.
>
> Events **EK** files are actually a collection of several file types. **ESP** files contain the science observation plans, **ESQ** contain spacecraft and instrument commands, **ENB** include notes for the spacecraft events and ground data logs. According to the NAIF, EK files are not commonly used.

There are a series of other file types that are sometimes needed. **FK** files contain definitions of reference frames and conversion relationships; **LSK** files include tabulations for leap seconds and other time conversions; **SCLK** include spacecraft clock to ground-based conversion information; and **DSK** files include (as necessary) shape models for the target (e.g. digital elevation model).

SPICE also includes a series of software modules, called **Toolkits**, which can be incorporated into the end user's software. These open-source and license-free modules are written in a variety of programming languages (e.g. C, MATLAB) and compatible with several different platforms and operating systems (PC/Windows, PC/Linux, Mac/OSX, Sun/Solaris). They can be downloaded from the NAIF website and include annotated source code, reference manuals, and user guides.

There are a host of applications enabled by the SPICE kernels and Toolkits. For example, it can be used not only to inform the user of the orientation and position of an image in the past, but can help to predict when the next opportunity for a restricted set of conditions will occur. These applications are extensively used by mission planners to develop operation sequences.

7.3 Case Study: Laboratory Tests of the Hapke Model

In planetary photometry, the Hapke model is ubiquitous. It has proven to be of great value for fitting phase curves (e.g. Mallama et al., 2002), for correcting spacecraft imagery to a standardized lighting and viewing geometry (e.g. Domingue et al. 2010, 2015; Sato et al., 2014), and for un-mixing the spectral contributions of intimate mixtures of materials to extract abundance estimates (e.g. Mustard and Pieters, 1987, 1989; Cruikshank et al., 2001). However, there have been few tests of its ability to extract information on the physical state of the regolith. One such test was conducted by Shepard and Helfenstein (2007).

A variety of test samples were chosen, prepared, and measured by one member of the team. The results were sent without context to the other member, and the best-fit Hapke model parameters were extracted. These were then examined by both members and compared to the properties of the samples as a check on the model.

7.3.1 Instrument System

The test used a goniometer (BUG lab, Section 7.1) with three independent degrees of motion: $0°–60°$ in incidence angle (source arm), $0°–80°$ in emission angle (detector arm), and $0°–180°$ in azimuthal rotation (sample stage). The light source was a 100 W quartz-halogen lamp with an active output stabilizing system. Its output was collimated, filtered using one of several broadband color interference filters, chopped, focused onto a fiber optic bundle, routed to the end of a 60-cm arm, and collimated again before being directed onto a sample dish below.

The sample dish was 56 mm in diameter and contained an optically thick (~1 cm) powdered material of choice. It was isolated from all axes of motion and did not move. The incident spot was ~20 mm at nadir and elongated into an ellipse on the sample as the source arm increased in incidence angle. Given the spot size and distance from the source, all incident rays were within 1° of the source arm angle of incidence.

The detector was a solid-state silicon broadband detector sensitive to wavelengths 400–2,500 nm, and was mounted at the end of the 80-cm detector arm. Its response curve was linear and the detector electronics were synchronized to the source-light chopper to minimize background noise and eliminate the effects of ambient light. Even so, all measurements were conducted in a darkened room. The entire space visible to the source and detector were flocked with black-out materials to minimize stray light.

All axes of motion were driven by stepper motors with 0.001° precision. Absolute pointing accuracy was estimated to be <0.5° and relative accuracy with respect to other axes was estimated to be <0.25°. Computer software moved the arms in a pre-arranged script of ~700 different combinations of incidence, emission, and azimuth angle designed to cover the entire scattering hemisphere in ~0.05 sr increments. Measurements were made at phase angles of 3°–140°, the lower limit based on the source arm occulting the sample, and the upper limit based on the largest spot size (elliptical) that would comfortably fit on the sample. All measurements were output to a text file.

7.3.2 Calibration

The detector output was in volts. To convert these into useful values, all measurements were divided by the detector output for a sample of Spectralon® measured just prior to all sample measurements. Because Spectralon® is known to have non-Lambertian properties at higher incidence and emission angles, the sample was only measured at a near-nadir geometry ($i = 0°$, $e = 10°$); dividing the reflectance of a sample at any geometry to that Spectralon® at this single geometry provided the radiance factor. Some goniometers measure the reflectance from a constant area *within* the illuminated spot; this goniometer measured the reflectance of the entire spot, even when elongated, so an additional cosine correction factor was necessary. The output of the lamp had an active feedback system and maintained lamp stability to <1% over the course of hours of observations per day. This was verified by comparing the raw detector voltages of the Spectralon® measurements at the beginning and end of a sample run.

Stray light is an issue in nearly all optical instruments. For this goniometer, the largest contributor was secondary scattering from the sample onto the surrounding instrumentation and into the detector. Flocking all surfaces except the sample minimized much of this, but some was still evident. Much of the remainder could be detected and removed by measurements at $i = 0°$, $e =$ constant, and rotating in azimuth. Checks on repeatability and accuracy were made by repeat measurements at the same and mirror azimuth geometry (same incident and emission angles but azimuths of, for example, $45°$ and $-45°$) during a sample run. All measurements were found to be repeatable to <5% RMS, and the absolute reflectance was conservatively estimated to be within 15% of true.

7.3.3 Observations and Results

Sixteen powdered samples were measured including fine-grained (~micron-scale) iron-rich clays, four different sizes of aluminum oxide abrasives (25–125 µm), course sands (300–500 µm), fine-grained (5–20 µm) dark powders, and fine-grained strongly colored powders (red iron oxide and green cobalt carbonate). Some samples were measured in multiple wavelengths to take advantage of strong color contrasts, and some were measured in various stages of compaction. All samples were prepared in the sample dish in the same way and were leveled so that there was no topography other than at the scale of the grain. Each sample was weighed and its bulk porosity estimated from its known grain density. In all, 29 bidirectional reflectance distribution function (BRDF) files were analyzed (Figure 7.2).

Observations were fit to four different versions of the Hapke model: (1) a two-term H-G particle phase function and only the shadow-hiding opposition effect (SHOE); a two-term H-G particle phase function with both SHOE and coherent-backscattering opposition effect (CBOE); (2) a three-term H-G function with only SHOE; (3) a three-term H-G function with both SHOE and CBOE.

The extracted single-scattering albedo (SSA) of each sample was found to be stable between model fits. An important finding, consistent with work of previous authors, was that the single-scattering albedo, w, was a function of packing state – a porous sample of material had a smaller w than a compacted sample of the same material. As Hapke (1999) had pointed out, this observation was a consequence of using a uniform media model (radiative transfer) to model a discrete media. A modification to the Hapke model attempts to account for this (Hapke, 2008).

Surprisingly, the authors found all samples inversions to require surface roughness, even though all samples had been smoothed at the sample container scale. However, there was evidence that the roughness coefficients were

216   Planetary Disk-Resolved Photometry

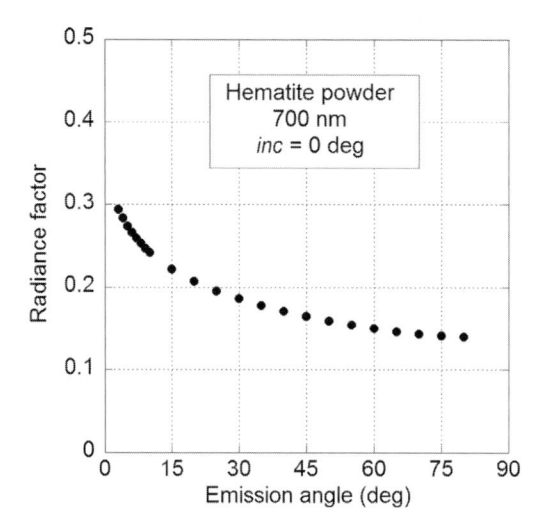

Figure 7.2A. Bidirectional reflectance of a laboratory sample of hematite (iron oxide) powder. Radiance factor at 700 nm is plotted versus emission angle. Incidence angle is constant at $i = 0°$. Note the strong opposition surge.
Credit: Michael K. Shepard.

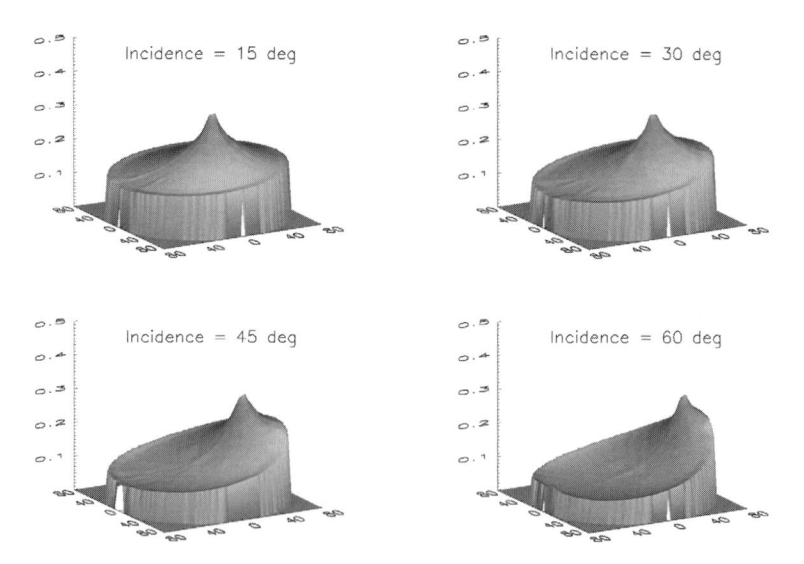

Figure 7.2B. Bidirectional reflectance of the same material, but for different incidence angles. Each figure shows a three-dimensional plot in which the vertical axis is the radiance factor and the radial distance from the center of the plot is the emission angle ($e = 0°$ in center, $e = 80°$ around edges).
Credit: Michael K. Shepard.

comparable to slopes at the particulate scale. There was a weak correlation between the roughness parameter and the opposition surge width, suggesting a link between those two parameters, and an inverse correlation between the roughness parameter and single-scattering albedo. This latter correlation was thought to be caused by the reduction of shadows at the smallest (and therefore roughest) scales of the surface. Finally, the authors found no apparent correlation between the measured bulk porosity and the opposition surge parameters. In summary, the work suggested that the retrieved parameters were due to complex interactions between different physical properties of a surface that were difficult to separate.

7.4 Case Study: Mercury and MESSENGER

An example that integrates both disk-integrated and disk-resolved data is summarized by Domingue et al. (2010). In this work, the authors use the extensive ground-based phase curves of Mallama et al. (2002) and Warell and Bergfors (2008; Chapter 6) in conjunction with disk-resolved measurements by two sensors on the Mercury MESSENGER spacecraft: the Mercury Dual Imaging System (MDIS) and the Mercury Atmospheric and Surface Composition Spectrometer (MASCS). The images were acquired during two flybys of the planet prior to orbital insertion in 2011.

One of the primary goals of the MESSENGER mission was to image 97% of the planet in eight color filters with a spatial resolution of 1 km or better. Mission constraints made it impossible to obtain uniform lighting and viewing geometry throughout the mission, so photometric correction to a standard lighting and viewing geometry was necessary to

1. Provide a seamless global mosaic of Mercury in eight colors; and to
2. Allow for direct comparison of eight-color spectra to laboratory spectra of potential compositional analogs.

The first goal is important for geological and geomorphological science; ideally, one would like imagery to be acquired under identical illumination and viewing conditions to ensure that observed differences are real and not simply photometric artifacts (Figure 7.3). Likewise, geological analysis needs compositional information derived from spectral observations. It is known that scattering properties are often wavelength dependent, so that spectra of the same rock type acquired under different illumination and viewing conditions could look quite different. For this reason, most lab spectra are acquired at standard incidence and emission angles. Again, understanding the surface

Figure 7.3. An oblique view of the center of Abedin crater on Mercury. Oblique images offer a different perspective of a planet and opportunities for off-nadir photometry. MDIS mosaic.
Credit: NASA/Johns Hopkins University Applied Physics Laboratory/Carnegie Institution of Washington. PIA19423.

scattering law and its wavelength dependence is important for normalizing spectra to a standard geometry, preferably one than can be directly compared with existing spectral libraries.

Temperatures at Mercury were challenging; proximity to the Sun increased solar radiation by a factor of 10 relative to the Earth, and Mercury itself is a 700 K blackbody. Keeping the electronics cool required a sunshade, innovative cooling systems, and strict spacecraft operation and monitoring protocols.

7.4.1 MESSENGER Camera

We focus on the results of the Mercury Dual Imaging System (MDIS). The specifications of this system are found in Hawkins et al. (2007, 2009) and references therein.

The MDIS was two camera systems in one: a monochrome narrow angle camera (NAC) and a wide-angle camera (WAC). Each camera used its own 1,024 × 1,024 frame-transfer CCD with 14 μm × 14 μm pixels. The NAC camera was an off-axis folded reflector with a 24.3 mm diameter primary mirror and a focal length of 550 mm (*f*/23). It used a single medium passband filter, centered on 750 nm. The WAC camera was a four-element refractor with a 7.8 mm objective aperture and a focal length of 78 mm (*f*/10). It included a 12-color filter wheel; 11 narrow-band filters covered passbands of diagnostic interest from 395 to 1040 nm, and the 12th used a clear filter for optical navigation.

7.4.2 Image Calibration

Most spacecraft cameras do not report brightness in stellar magnitudes. Instead, they report in spectral radiance units, e.g. W m^{-2} μm^{-1} sr^{-1}. Calibration of a spacecraft camera is a large, exacting task and well beyond the scope of this book. The bulk of the calibration is done pre-flight in a controlled setting, but it is virtually impossible to simulate the flight environment on the ground, so there are often in-flight calibration tests designed to check the early ground-based work.

7.4.2.1 Calibration

For a given MDIS pixel of value DN at position (x, y) on an image, the radiance, L, is given by the following (Domingue et al., 2010)

$$L(x,y,f,T,t,bm) = \frac{Lin[DN(x,y,f,T,t,bm,MET) - Dk(x,y,f,T,t,bm,MET) - Sm(x,y,t,bm)]}{Flat(x,y,f,bm) \times t \times RESP(f,bm,T)}$$

7.1

where DN is the raw pixel value, Dk is the dark current function, Sm is a function to correct for frame-transfer smear, Lin is a function that corrects for non-linear behavior in the CCD, $Flat$ is the flat field, and $RESP$ is a responsivity function – a calibration term that converts the corrected DN into radiance units (analogous to the zero-point function for telescopic calibration). The functional terms include the filter used, f; the temperature of the camera, T; the exposure time in milliseconds, t; the binning mode (single pixel, 2 × 2, etc.), bm; and the mission elapsed time (seconds since launch), MET.

One can see many of the same elements in this equation as discussed for the telescope calibration. There is a dark-frame correction, but it can be (and often is) a function of time since launch. There is a flat-field frame, but as described by Howell et al. (2007, 2009), there were two major issues for the MDIS. First, the original laboratory flat-field was plagued by scattered light from a thermal radiator while in the environment chamber, so it was taken out of the environment chamber and the flat-field was obtained at room temperature. This led to artifacts from the higher temperatures. The WAC also suffered from flat-field "donut" shadows. These resulted from tiny particulate contamination on the CCD cover glass near the focal plane. These moved after launch, requiring an additional in-flight flat-frame using an onboard calibration target; these suffered from scattered light contamination, so the final flat-field frame used a combination of elements from both the ground and in-flight versions.

Even the best CCDs are rarely linear up to full well depth capacity, so a correction term is required, especially for brighter targets where the non-linearity is likely to be most conspicuous. Because the MDIS did not contain a mechanical shutter, there is some smear associated with the frame-transfer of the image prior to readout, and this is accounted for in the *Sm* function.

There was also a change in the MDIS calibration performance after May 2011. This event had no effect on the results discussed here, but had to be considered in subsequent papers (Domingue et al., 2015). The change was thought to have been triggered by a close approach to a particularly hot region of Mercury (Keller et al., 2013).

7.4.2.2 Scattered Light

Not included in this calibration function are corrections for scattered light – light either from sources outside the *FOV* or inadvertent multiple reflections from targets within the *FOV*. Many of the corrections must be empirically determined in-flight.

For the NAC, approximately 2% of the measured light was estimated to be from sources outside the *FOV* for uniform sources. In other words, while looking at a particular feature on Mercury, light from areas around that feature, but not visible within the frame, were contributing extra light. This was thought to be a result of the off-axis design of the NAC (Hawkins et al., 2007, 2009).

The WAC exhibited more scattered light than the NAC. When looking at the limb of Mercury, one expects a sharp cutoff in DN. Instead, it was often found that there was a gradient of scattered light creating a faint but diffuse glow up to several degrees away from the limb. And scattered light in the WAC was found to have a wavelength dependence that unfortunately mimicked realistic spectral features (Hawkins et al., 2007, 2009). All of these effects required quantification under a variety of conditions for proper accounting.

7.4.2.3 Radiance Factor

As noted in Chapter 5, radiance by itself is rarely of value. This value must be normalized to some measure of reflectance. The most commonly used measure in mission imagery is probably radiance factor, r_f, (often abbreviated *I/F* in planetary literature), defined as the radiance from the target relative to that from a flat Lambertian reflector of the same area, at the same distance from the Sun and viewer, under normal (vertical) illumination. For the MDIS targets, Domingue et al. (2010) give the following conversion

$$r_f = \pi \frac{L(f)}{F(f)} \left(\frac{d_M}{d_{AU}}\right)^2 \qquad 7.2$$

where $L(f)$ is the radiance measured from Mercury using filter f, $F(f)$ is the solar irradiance through filter f at a distance of 1 AU, d_M is Mercury's distance from the Sun at the time of the measurement, and d_{AU} is the distance of 1 AU (both distances should use the same units). Inspection of this formula will show it to be identical to the definition of radiance factor (Eq. 5.10) with the inverse square law modification to account for the difference in distance between Mercury and the solar standard (F) at 1 AU.

7.4.2.4 Optical Distortion and Alignment

Even excellent optical systems have some distortion. One of the tasks during preflight calibration is modeling of this to ensure all pixels are as free from aberration as possible. Each camera will have an associated camera model; in the case of the MDIS, these were radially symmetric polynomials to correct for minor distortion near the frame edges (Hawkins et al., 2007, 2009).

It is often desirable to compare images from different instruments, for example to use spectral information from one to overlay upon another, or to create mosaic images. Understanding their relative image alignment is necessary to do this properly. In the case of the MDIS, the WAC provided context and the NAC provided detail. Empirical alignment functions were generated in-flight by comparing star fields or other well-controlled imaging targets.

7.4.3 Comparison with Ground-Based Disk-Integrated Observations

During the two flybys (M1 in January 2008 and M2 in October 2008), MDIS obtained several whole-disk images that were used to generate radiance factors that could be compared directly with the extensive ground-based whole-disk photometry that had accumulated over the past few decades (Mallama et al., 2002; Warell and Bergfors, 2008; Chapter 6).

Background DN were averaged from regions away from the planetary disk and used with the dark current correction to remove background noise. All pixels containing the image of Mercury were then averaged to obtain a disk-integrated radiance factor. Each image provided one value at one phase angle in one filter. These were directly compared with previous phase curves by converting the reduced magnitude phase curve into radiance factor phase curve using the following relationship:

$$r_f(\alpha) = \left(\frac{\pi}{\Omega}\right) 10^{-0.4(m - m_{sun})} \quad \quad 7.3$$

Where Ω is the solid angle of Mercury at 1 AU distance, m is the reduced magnitude of Mercury, and m_{sun} is the magnitude of the Sun in the appropriate filter. This process was not completely straightforward because the filter sets used by MDIS did not correspond directly with the Johnsons-Cousins standards. Despite these issues, the two data sets were found to be generally consistent.

7.4.4 Results

To develop a global photometric function for Mercury in eight colors, Domingue et al. (2010) combined disk-resolved images from the two flybys with disk-integrated observations, both spacecraft and ground-based, to generate a master photometric data set for analysis. They chose to use one of the earlier, simpler forms of the Hapke model to fit this data set; they included only the single-scattering albedo (w), shadow-hiding opposition surge parameters (width, h, and magnitude, B_0), the Hapke roughness parameter (theta-bar), and a two-term (b, c) Henyey-Greenstein function for the phase function.

The disk-resolved data were taken from an equatorial region near longitude 120° and covered phase angles from 51° to 120°. Unfortunately, these phase angles provide no constraints on the opposition surge and limited constraints on the phase function and roughness parameters; these parameters require phase angles of 0°–20°, and >120°, respectively. There is also a noted coupling between single-scattering albedo and the Hapke roughness parameter that requires some resolution (Helfenstein and Veverka, 1989). As a result, it was decided to fit the model in a multi-stage process.

The V-filter disk-integrated data of Mallama et al. (2002) cover phase angles from 2° to 170°, the largest range of any ground-based data set. These were combined with the 560 nm MDIS disk-integrated observations and fit with a Hapke model to derive the opposition surge width, $h = 0.075$, and magnitude, $B_0 = 2.3$, of the shadow-hiding opposition effect. Because it is widely assumed that the SHOE is wavelength independent, these values were adopted for all subsequent fits in all wavelengths.

In the next step, the disk-resolved data were modeled in all wavelengths. The opposition surge parameters were fixed by the previous step, but all other parameters were allowed to float. The results of these fits showed the Hapke roughness parameter to hover around a common value with a median of 9°. As with the SHOE, this parameter is widely believed to be wavelength independent and was thus fixed for all subsequent fits.

In the final step, the disk-integrated and disk-resolved data sets in all wavelengths were fit. The opposition surge and roughness parameters were now

7.4 Case Study: Mercury and MESSENGER

fixed from the previous steps. The remaining parameters – single-scattering albedo, w, and phase function terms, b and c – are all wavelength dependent and differ from band to band.

Subsequent to orbit insertion in 2011, the MESSENGER team obtained thousands of additional MDIS images, some of which were targeted specifically for additional photometric calibration. Application of the preliminary photometric corrections described previously were found to leave seams between images. Despite considerable effort and multiple attempts, the photometric correction could not be made completely seamless, and was especially problematic for images where angles (i, e, or α) exceeded 50°. The results of that effort are outlined in Domingue et al. (2011, 2015; Figure 7.4).

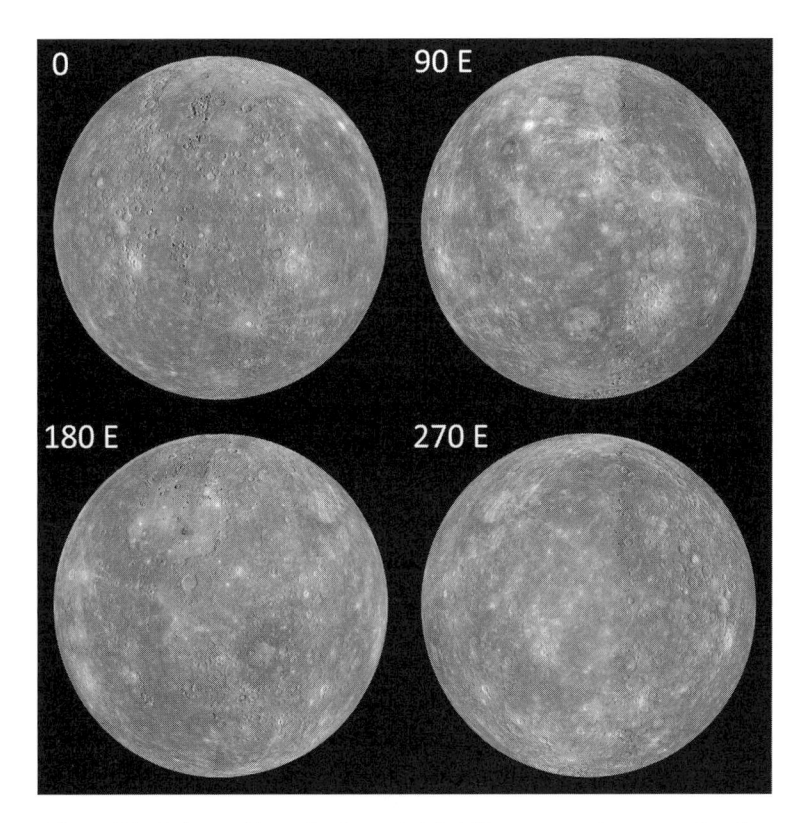

Figure 7.4. Orthographic projections of global Mercury mosaics centered at the equator and meridians noted in the top left corner. Both MDIS NAC and WAC were used in these mosaics. Resolution 2.5 km/pixel.
Credit: Michael K. Shepard and NASA/Johns Hopkins University Applied Physics Laboratory/Carnegie Institution of Washington. PIA15160–15163.

7.5 Case Study: The Moon and Clementine

The Clementine spacecraft was launched on January 25, 1994, entered lunar orbit on February 19, and mapped the entire Moon until May 3. It was a joint NASA and US Department of Defense mission with a major mission goal of testing new spacecraft hardware emphasizing light weight and low cost. Its second mission goal was to conduct a scientific investigation of the Moon and later, the asteroid 1620 Geographos; unfortunately, hardware and software issues precluded the Geographos flyby. Clementine carried seven different instruments, including cameras to image the Moon in the ultraviolet and visible spectrum (UVVIS) and the near-infrared (NIR), and a laser ranging system (LIDAR) for topographic mapping.

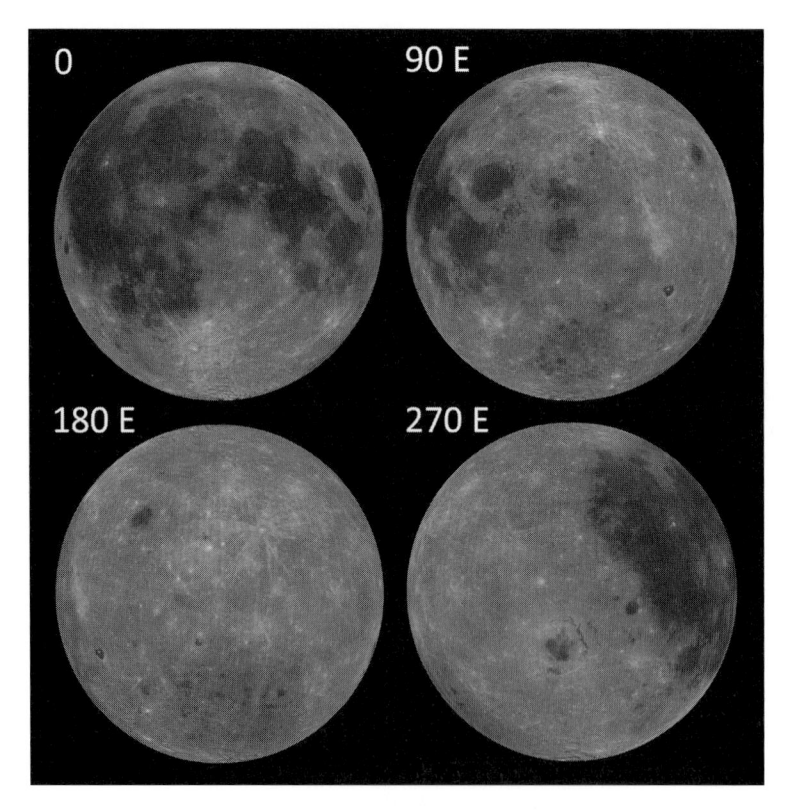

Figure 7.5. Orthographic projections of global lunar mosaics centered at the equator and meridians noted in the top left corner. Seams between image strips indicate an imperfect photometric correction. UVVIS camera, 750 nm.
Credit: Michael K. Shepard and NASA/JPL/USGS. PIA00302–00305.

7.5 Case Study: The Moon and Clementine

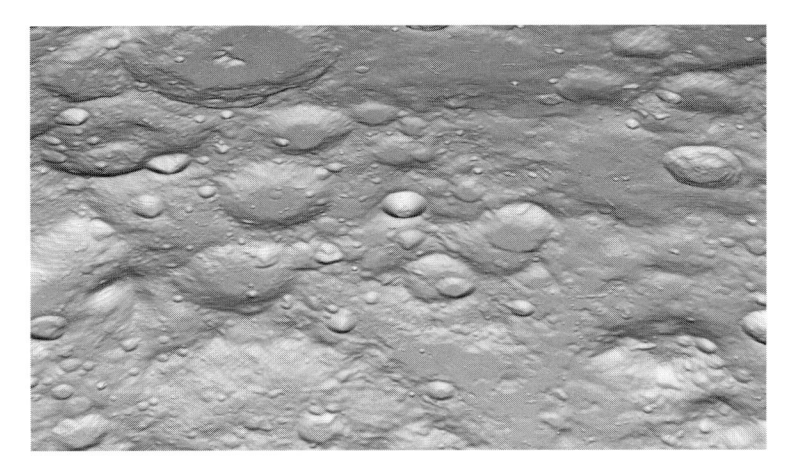

Figure 7.6. A perspective view of the south pole of the Moon. The bowl-shaped crater at the center is Shackleton crater. This image was derived from Clementine UVVIS data overlain on topographic data from the Lunar Reconnaissance Orbiter (LRO) Lunar Orbiter Laser Altimeter (LOLA). At this stage, no shadowing or realistic lighting has been added.
Credit: NASA/Goddard Space Flight Center Scientific Visualization Studio. http://svs.gsfc.nasa.gov/3731

The Clementine data were the first major global data set of the Moon since the Lunar Orbiter missions of the 1960s. The multicolor digital data set was ideal for mapping the entire Moon in a systematic fashion and investigating any number of problems (Figure 7.5). Many found pairing the images with the topographic data to be an especially powerful visualization tool (Figure 7.6).

Here, we focus on the properties and results of the UV/Visible camera. A number of investigators used these data to study phase reddening (Hillier et al., 1999; Kreslavsky et al., 2000), the lunar disk function (Kreslavsky et al., 2000), photometric oddities (Kreslavsky and Shkuratov 2003), and especially the opposition surge and its causes (Buratti et al., 1996; Hillier et al., 1999; Shkuratov et al., 1999; Kreslavsky et al., 2001). Here we focus on Hillier et al. (1999) because it outlines the effort to create photometric corrections for the lunar surface and offers a case study for understanding phase reddening and the opposition surge phenomenon.

7.5.1 Clementine UV/Visible Camera (UVVIS)

7.5.1.1 Camera and Optics

The optics for the UVVIS consisted of a catadioptric telescope with an aperture of 46 mm and focal ratio $f/1.96$. It used a Thomson frame-transfer CCD with

288 × 384 pixels, each 23 μm × 23 μm. The CCD was sensitive to wavelengths ranging from 250 to 1000 nm and used one of six filters when imaging. For convenience, the narrow-band filters were given a letter designation: A (415 ± 20 nm), B (750 ± 5 nm), C (900 ± 10 nm), D (950 ± 15 nm), and E (1000 ± 15 nm). There was also a broadband channel (400–950 nm). The electronics had three gain states of 100, 350, and 1000 e$^-$/DN and digitized all output using an 8-bit A/D converter.

7.5.1.2 Orbit

The Clementine orbit was an elliptical polar orbit (pole-to-pole, perpendicular to the equator), with periapsis of 400 km above the lunar surface and apoapsis of 8,300 km. An orbit took approximately 5 h. In the first science mapping phase, the periapsis was centered over −30° latitude so that the southern hemisphere to −60° could be imaged at high resolution. This phase lasted one month, then the orbit was rotated so that periapsis was at +30° latitude and the second month-long mapping phase took place, covering the northern hemisphere to +60° latitude.

In both science mapping phases, imaging took place for 80 minutes around closest approach, and data downlink took place for a little more than 2 h at apoapsis. The orbit was such that the Moon rotated ~3° underneath per orbit, so that the imaging sequences were long, overlapping strips, and ~120 orbits were sufficient to imaging the entire surface within the latitude restrictions.

7.5.2 Image Calibration

A summary of the calibration steps for the Clementine UVVIS imager is given by McEwen (1996) and includes corrections for dark current, non-linearity, frame-transfer (smearing), flat-fields, and distance (from the Sun) normalization.

Observations could be either radiometrically or photometrically calibrated. Radiometric calibration was based on in-flight observations of the standard star Vega. It and a number of other standard star's output (AB magnitudes) are well documented[2] and can be converted into absolute spectral radiance units (Chapter 2), allowing engineers to convolve the response of any spacecraft filter set and establish absolute calibration factors.

The photometric calibration of the UVVIS observations into reflectance (radiance coefficient, aka reflectance factor) was originally based on comparison of laboratory spectra of Apollo 16 regolith and Clementine observations,

---

[2] A list of AB magnitude standard stars can be found at the European Southern Observatory website: www.eso.org.

or **ground-truth**, of the Apollo 16 landing site. The calibration was later found to be too high by a factor of ~2 and subsequently corrected.

7.5.3 Comparison with Ground-Truth

The goal of photometric calibration is to ensure that any set of observations, at any illumination and viewing geometry, can be corrected to some standard illumination and viewing geometry. Some applications of this include stitching together images to make a seamless mosaic, looking for changes between images of the same place but taken at different times under different illumination conditions, or for comparison with laboratory spectral data to extract compositional information. Fortunately for Clementine, there is a large history of ground-based photometric observations and characterization of the Moon.

The first attempt to calibrate the Clementine imagery used what should have been the gold standard: a sample of lunar regolith taken from the landing site of Apollo 16 (sample 62231). This sample was well characterized, was similar to other samples taken from the landing area, and was measured using the NASA Reflectance Laboratory (RELAB) spectrometer at Brown University (Pieters, 1999).

The RELAB spectrometer measured the Apollo sample from 400 to 2,500 nm at a standard illumination and viewing geometry, $i = 30°$, $e = 0°$, and calculated the radiance coefficient (reflectance factor) by comparing it to a sample of Spectralon® measured at the same geometry. The spectrum was convolved with the Clementine filter response curves to generate the equivalent radiance coefficients of the sample in each filter.

To compare the laboratory data with the spacecraft observations, average DNs were extracted from a homogeneous 12×33-pixel area west of the Apollo 16 landing site in the Clementine image data (in each filter), and corrected to the same standard illumination and viewing geometry ($i = 30°$, $e = 0°$) using an a priori lunar photometric model (McEwen 1996). Finally, empirical multiplicative correction factors were calculated for each filter to make the average DN of the reference site equal the radiance coefficient of the lab sample. These correction factors then became part of the Clementine calibration pipeline.

It was later noted that the calibration appeared to make the Moon too bright (Blewett et al., 1997) by a factor of ~2. Some suggestions for this discrepancy included differences in the packing state or surface roughness between in situ and lab samples, or possibly something as straightforward as an incorrect transformation of the Clementine observation and viewing geometry to the standard.

Hillier et al. (1999) corrected the calibration issue by using the ground-based observations of the Moon by Shorthill et al. (1969). In this data set, the authors observed a large number of regions over an entire lunation ensuring a wide range of illumination conditions. Hillier et al. extracted the reflectances of 15 representative areas from this catalog that were similarly observed by Clementine and made these the calibration standard for the spacecraft observations.

7.5.4 Results

Hillier et al. (1999) used the Clementine imagery to investigate two major photometric properties of the lunar surface: phase reddening and the opposition surge. They began by extracting radiance factors (they called these I/F) in all filters from both maria and highland areas at a wide range of phase angles. The data from each filter were then fit with an empirical model of the form

$$r_f = \frac{\mu_0}{\mu_0 + \mu} P(\alpha) \qquad 7.4$$

where $P(\alpha)$ is an empirical phase function that includes albedo and other terms.

To determine if phase reddening (see Chapter 6) was present, they divided the phase curves of all filters by $P(\alpha)$ of the B filter (750 nm). For the maria, phase reddening was evident when the B filter was compared with the violet A filter but not the 1 μm E filter. In the highlands, however, phase reddening was visible in *both* the A and E filters, although there was evidence of phase bluing in the highlands A/B ratio for phase angles $\alpha > 70°$.

There is debate about the causes of the opposition surge; most agree that shadow hiding is a dominating mechanism. However, for some surfaces, especially brighter ones, there is evidence for a much narrower opposition surge ($\alpha < 2°$) attributed to coherent backscatter. Because Clementine orbited the Moon, it was capable of observing at phase angles smaller than possible from Earth, although still limited by the finite angular diameter of the Sun.

To study the opposition surge, Hillier et al. (1999) focused on phase angle data $\alpha \leq 10°$ and employed a full Hapke model with a two-term H-G phase function, roughness parameters, and shadow hiding and coherent backscatter opposition surge widths and magnitudes. They found evidence for a narrow (1°–2° width) coherent backscatter component (~15% of the shadow hiding enhancement) in the highlands region, but no similar component in the maria. There was also no evidence for a wavelength dependence in the opposition surge observations, suggesting a dominance of the shadow-hiding mechanism. This was not unexpected: coherent backscattering is a multiple-scattering phenomenon, so it is of little importance in darker surfaces like the maria and of limited importance in the slightly brighter highlands.

Appendix 1
Problems

Chapter 1

1.1 If a linear polarizer is used to filter light, the output intensity, I, is proportional to the input intensity, I_0 by Malus' law:

$$I = I_0 \cos^2\theta$$

where θ is the angle between the polarization axis of the filter and the light's original polarization orientation. For unpolarized light, the mean value of $\cos^2\theta = 0.5$ so that a perfect polarizer cuts the initial light by 50%. A second linear polarizing filter in the path would cut the light down by an additional factor, depending upon its orientation with the first filter.

Compare the light from a star A and a planet B, both point sources. The star is viewed with no filtering by the unaided eye, while the planet is viewed simultaneously, but only after filtering with perfect twin crossed-polarizers. The two appear to be the same brightness when the angle between the two polarizers is 60°. What is the difference in apparent magnitude of A and B?

1.2 If we were instead comparing the brightness of these two stars with identical telescopes, by what factor should telescope B be stopped down to make the two images identical in apparent brightness?

Chapter 2

2.1 What is the angular distance between two objects: A is at *RA/DEC* (25°, −10°) and B is at (90°, +30°)?

2.2 What is the maximum phase angle for observing Neptune's moon Triton from Earth?

2.3 A surface is illuminated by the Sun at $i = 60°$ at an azimuth angle of 75° and is viewed by a sensor at an emission angle of $e = 30°$ and azimuth angle of 120°. What is the phase angle?

2.4 By the time this book is published, at least one additional leap second will have been added. Find the current difference between Atomic Time and *UTC*.

2.5 Jupiter is at opposition and 4.2 AU from Earth when shadow of Europa transits across Jupiter. You record the first contact of the shadow at 02:37:15 *UTC*. What is the travel time and light-corrected time of this event?

2.6 What is the solid-angle subtended by the Moon from the *surface* of the Earth?

2.7 If two stars differ by 0.1 mag, what is their approximate difference in irradiance?

2.8 Venus has an apparent magnitude of $V = -4.17$. It is at a phase angle of 72.3°, 0.719 AU from the Sun and 0.960 AU from Earth. What is its reduced magnitude?

Chapter 3

3.1 A compound telescope has a primary mirror of diameter $D = 0.5$ m and focal length $fl = 1.1$ m. What is its focal ratio? When paired with a hyperbolic secondary mirror, the focal ratio is extended to $f/12$. What is the effective focal length?

3.2 What is the theoretical resolution of this telescope in the visual wavelengths?

3.3 A CCD camera is to be paired with this telescope at a site where the best seeing is 1.0″ and a sampling parameter of 2.5 is desired. What pixel size does this correspond to?

3.4 The camera chosen has 4500 × 3600 pixels of 6 μm size (square). What plate-scale does this correspond to? What is the best operational mode when seeing is optimal?

3.5 Assuming the fill factor of the chip is 100%, what is the physical size of the chip and the apparent field of view?

3.6 Each pixel has a full well capacity of 40,000 e⁻. What is the best *SNR* achievable when in 2 × 2 binning mode?

3.7 The readout noise for this camera is 8 e⁻ rms. Assuming 2 × 2 binning, what is the dynamic range? In decibels? What number of bits are needed to fully digitize the output?

Chapter 4

4.1 Opticians use high-index material for prescription eyeglasses to make them thinner. For a lens with $n = 1.74$, what fraction of incident light is reflected (assuming no special coatings)?

4.2 The complex index of refraction of silica glass at 550 nm is $1.5 + i10^{-7}$. What is the expected attenuation of a beam of light traveling through a 1 m thickness of this material?

4.3 Oceanographers frequently measure the light penetration profile of a water column and report an attenuation coefficient, κ (m^{-1}), which includes the effects of suspended particulates, dissolved material, and the intrinsic absorption of water. For a region of San Francisco Bay, a recent study reported $\kappa = 1.2$. Use Beers law to determine the depth at which light is 1% of the surface value.

4.4 What is the depth of water in #3 corresponding to an optical depth $\tau = 1$?

4.5 Prove that Equation 4.14 holds (i.e. that $P(\alpha)$ is normalized) when $P(\alpha) = 1$ (Equation 4.18).

Chapter 5

5.1 Demonstrate that the Lommel–Seeliger function has a uniform (flat) disk function at opposition.

5.2 Derive the quantities shown in Table 5.1 for the Minnaert function.

5.3 Show that Equation 5.68 is the solution of Equation 5.65.

5.4 Show that Equation 5.74 is the solution of Equation 5.73.

Chapter 6

6.1 What is the physical (or geometric) albedo for asteroid 21 Lutetia given the following: $H = 7.35$, $D_{eff} = 96$ km?

6.2 In the following table there are 20 reduced visual magnitudes for an asteroid and their respective phases. Make a plot of the data and determine the following:
   a. Phase coefficient, β, and $V(1, 0)$ (Section 6.4.2).
   b. H and G
   c. Solve Equation 6.13 for $V(1, 0)$, A, and B. What is $V(0)$?

α (°)	$V(1, \alpha)$
0.9	6.05
1.4	6.08
1.6	6.09
2.5	6.15
3.2	6.19
4.0	6.24
4.6	6.27
5.7	6.33
6.5	6.37
8.0	6.39
8.6	6.43
9.3	6.48
11.5	6.52
12.0	6.57
13.5	6.62
14.1	6.61
14.3	6.62
16.0	6.70
16.9	6.70
18.1	6.78

6.3 Given the best fit phase coefficients in Problem #2, estimate the geometric albedo of this asteroid using Equation 6.25.

6.4 Assume this geometric albedo is good to ±0.02. Estimate the size of this asteroid with uncertainties.

6.5 What polarization phase slope and P_{min} would be expected if we assume the geometric albedo in 6.3 is correct?

6.6 For the ground-based telescope/CCD in the Mallama et al. Mercury study (Section 6.6), find the following:
   a. the plate scale or *IFOV*
   b. the theoretical resolution of the stopped-down instrument
   c. the sampling parameter?
   d. the field of view assuming 100% fill factor

6.7 What is the maximum and minimum expected angular size of Mercury as seen from Earth? Assume a diameter of 4,900 km, and circular orbit of 0.4 AU. How does this compare with the plate scale and *FOV* of the telescope?

6.8 Derive Equation 6.18 using the definition of phase integral and magnitude. (Hint: Equation 6.24 may help you see where to start.)

Chapter 7

7.1 Assuming the MESSENGER MDIS CCD has 100% fill factor, compute its dimensions and use that to determine the field of view of both (NAC and WAC) cameras. In addition, calculate the plate scale for each camera.

7.2 Compare the theoretical resolution of each MDIS camera optics with their *IFOV*.

7.3 What is the *FOV* and *IFOV* of the Clementine UVVIS camera?

7.4 Assuming a nominal Clementine orbital altitude of 400 km, what is the cross-track width of the images? What is the nominal resolution per pixel?

7.5 Derive Equation 7.3.

7.6 (Open-ended problem). You are on the proposal team for a mission to study Europa. Because of the harsh radiation around Europa, the mission will orbit Jupiter and flyby Europa for imaging. Assume a typical flyby range of 5,000 km with some as close as 50 km. The mission specifications require mapping the entire surface to resolutions of 50 m. Design a possible narrow- and wide-angle camera system. Consider the following factors:
   a. amount of sunlight available at Europa and its albedo.
   b. the speed of the encounter and methods to minimize blurring.
   c. type of imaging system – array, pushbroom, or pushframe.
   d. camera focal length, theoretical resolution, *IFOV*, and field of view.
      Afterward, look up previous or current design proposals for this kind of system.

References

Airy, G.B. 1835. On the diffraction of an object-glass with circular aperture. *Trans. Camb. Philos. Soc.*, **5**, 283–291.

Akimov, L.A. 1979. On the brightness distribution across the lunar disk and planets. *Astron. Zh.*, **56**, 412–418 (in Russian).

Akimov, L.A. 1988. Light reflection by the Moon. *I. Kinemat. Phys. Celest. Bodies*, **4** (1), 3–10 (in Russian).

Arago, F. and Fresnel, A. 1819. On the action of rays of polarized light upon each other. In H. Crew, ed., *The Wave Theory of Light – Memoirs by Huygens, Young and Fresnel*. American Book Company: New York, NY. 145–155.

Barabashev, N.P. 1922. Bestimmung del' Erd albedo und des Refiexionsgesetzes für die Oberflache del' Mondmeere Theorie del' Rillen. *Astron. Nachr.*, **217**, 445–452.

Baranne, A. and Launay, F. 1997. Cassegrain: A famous unknown of instrumental astronomy. *J. Optics*, **28** (4), 158–172.

Bartholin, E. 1670. An account of sundry experiments, made and communicated by that learned mathematician, Dr. Erasm. Bartolin, upon a chrystal-like body sent out of Iceland. *Philos. Trans.* (1665–1678), **5**, 2039–2048.

Bell, L. 1917. The physical interpretation of albedo. *Astrophys. J.*, **45**, 1.

Belskaya, I. and Shevchenko, V.G. 2000. Opposition effect of asteroids. *Icarus*, **147**, 94–105.

Belskaya, I., Cellino, A., Gil-Hutton, R., Muinonen, K., and Shkuratov, Y. 2015. Asteroid Polarimetry. In P. Michel, F.E. DeMeo, and W.F. Bottke, eds., *Asteroids IV*. University of Arizona Press: Tucson, AZ. 151–164

Berry, E.M. 1923. Diffuse reflection of light from a matt surface. *J. Opt. Soc. Am.*, **7**, 627–633.

Bessell, M.S. 1979. UBVRI photometry. II – The Cousins VRI system, its temperature and absolute flux calibration, and relevance for two-dimensional photometry. *Astron. Soc. Pac.*, **91**, 589–607.

Bessell, M.S. 1990. UBVRI passbands. *Publ. Astron. Soc. Pac.*, **102**, 1181–1199.

Bessell, M.S. 2005. Standard photometric systems. *Ann. Rev. Astron. Astrophys.*, **43**, 293–336.

Bhandari, A., Hamre, B., Frette, O., et al. 2011. Bidirectional reflectance distribution function of Spectralon white reflectance standard illuminated by incoherent unpolarized and plane-polarized light. *Appl. Opt.*, **50**, 2431–2442.

Birnbaum, M.M. 1982. Voyager spacecraft images of Jupiter and Saturn. *Appl. Opt.*, **21**, 214–227.

Blewett, D.T., Lucey, P.G., Hawke, B.R., Ling, G.G., and Robinson, M.S. 1997. A comparison of mercurian reflectance and spectral quantities with those of the Moon. *Icarus*, **129**, 217–231.

Bockelee-Morvan, D., Gil-Hutton, R., Hestroffer, D., et al. 2015. Physical Study of Comets and Minor Planets, Legacy Report-2015, Division F Commission 15. Trans. IAU, Vol XXIXA, Proc. XXVIII IAU General Assembly, August 2012.

Bohren, C.F. and Huffman, D.R. 1983. *Absorption and Scattering of Light by Small Particles*. John Wiley & Sons: New York, NY.

Bottke Jr., W.F., Vokrouhlicky, D., Rubincam, D.P., and Nesvorny, D. 2006. The Yarkovsky and YORP effects: Implications for asteroid dynamics. *Ann. Rev. Earth Planet. Sci.*, **34**, 157–191.

Bouguer, P. 1729. *Essai d'Optique, sur la gradation de la lumiere*. Claude Jombert: Paris, France.

Bowell, E. and Lumme, K. 1979. Colorimetry and magnitudes of asteroids. In T. Gehrels, ed., *Asteroids*. University of Arizona Press: Tucson, AZ. 132–169.

Bowell, E., Hapke, B., Domingue, D., et al. 1989. Application of photometric models to asteroids. In R.P. Binzel, T. Gehrels, and M.S. Matthews, eds., *Asteroids II*. University of Arizona Press: Tucson, AZ. 524–556.

Bowman-Cisneros, E. and Eliason, E. 2011. *LROC RDR Data Products, Software Interface Specification*, v. 1.4. University of Arizona Press: Tucson, AZ.

Buratti, B.J., Hiller, J.K., and Wang, M. 1996. The lunar oppositions surge: Observation by Clementine. *Icarus*, **124**, 490–499.

Bureau International des Poids et Mesures (BIPM). 1979. 16th Conference Generale des Poids et Mesures (CGPM) Resolution 3. www.bipm.org/en/CGPM/db/16/3/.

Cellino, A., Belskaya, I.N., Bendjoya, Ph., et al. 2006. The strange polarimetric behavior of Asteroid (234) Barbara. *Icarus*, **180**, 565–567.

Cellino, A., Bagnulo, S., and Gil-Hutton, R. 2015. On the calibration of the relation between geometric albedo and polarimetric properties for the asteroids. *Mon. Not. R. Astron. Soc.*, **451**, 3473–3488.

Cellino, A., Bagnulo, S., Gil-Hutton, R., et al. 2016. A polarimetric study of asteroids: Fitting phase-polarization curves. *Mon. Not. R. Astron. Soc.*, **455**, 2091–2200.

Chandrasekhar, S. 1960. *Radiative Transfer*. Dover: NY.

Cheng, A.F., Weaver, H.A., Conard, S.J., et al. 2008. Long-range reconnaissance imager on new horizons. *Space Sci. Rev.*, **140**, 189–215.

Clegg, R.N., Jolliff, B.L., Robinson, M.S., Hapke, B.W., and Plescia, J.B. 2014. Effects of rocket exhaust on lunar soil reflectance properties. *Icarus*, **227**, 176–194.

Cord, A.M., Pinet, P.C., Daydou, Y., and Chevrel, S.D. 2003. Planetary regolith surface analogs: Optimized determination of Hapke parameters using multi-angular spectro-imaging laboratory data. *Icarus*, **165**, 414–427.

Cord, A.M., Pinet, P.C., Daydou, Y., and Chevrel, S.D. 2005. Experimental determination of the surface photometric contribution in the spectral reflectance deconvolution processes for a simulated martian crater-like regolithic target. *Icarus*, **175**, 78–91.

Cox, A.N. (ed.) 2000. *Allen's Astrophysical Quantities*. AIP Press, Springer-Verlag: New York, NY.

Cruikshank, D., Dalle Ore, C.M., Geballe, T.R., et al. 2001. Constraints on the composition of Trojan Asteroid 624 Hektor. *Icarus*, **153**, 348–360.

Danjon, A. 1949. Photometrie et colorimetrie des planetes Mercure et Venus. *Bull. Astron.*, **14**, 315–345.

Danjon, A. 1953. Correction to Danjon 1949. *Bull. Astron.*, **17**, 363.

Dershowitz, N. and Reingold, E.M. 2008. *Calendrical Calculations*. 3rd Ed. Cambridge University Press: Cambridge, UK.

DiLaura, D.L. 2001. *Lambert's Photometry, or, On the Measure and Gradations of Light, Colors and Shade*. Translation from the Latin with Introductory monograph and notes. Illuminating Engineering Society of North America: New York, NY.

DiLaura, D.L. 2006. *A History of Light and Lighting: In Celebration of the Centenary of the Illuminating Engineering Society of North America*. Illuminating Engineering Society of North America: New York, NY.

Dollfus, A. and Geake, J.E. 1977. Polarimetric and photometric studies of lunar samples. *Philos. Trans. Roy. Soc. Lond.*, **285**, 397–402.

Dollfus, A., Wolff, M., Geake, J.E., Lupishko, D.F., and Dougherty, L.M. 1989. Photopolarimetry of asteroids. In R.P. Binzel, T. Gehrels, and M.S. Matthews, eds., *Asteroids II*. University of Arizona Press: Tucson, AZ. 594–615.

Domingue, D.L., Vilas, F., Holsclaw, G.M., et al. 2010. Whole-disk spectrophotometric properties of Mercury: Synthesis of MESSENGER and ground-based observations. *Icarus*, **209**, 101–124.

Domingue, D.L., Murchie, S.L., Denevi, B.W., et al. 2011. Photometric correction of Mercury's global color mosaic. *Planet. Space Sci.*, **59**, 1873–1887.

Domingue, D.L., Murchie, S.L., Denevi, B.W., Ernst, C.M., and Chabot, N.L. 2015. Mercury's global color mosaic: An update from MESSENGER's orbital observations. *Icarus*, **257**, 477–488.

Durech, J., Sidorin, V., and Kaasalainen, M. 2010. DAMIT: A database of asteroid models. *Astron. Astrophys.*, **513**, DOI: 10.1051/0004-6361/200912693.

Evans, J.V. and Pettengill, G.H. 1963. The scattering behavior of the Moon at wavelengths of 3.6, 68, and 784 centimeters. *J. Geophys. Res.*, **68**, 423–447.

Fairbairn, M. 2004. On reflectance laws and the theory of planetary photometry. *J. Roy. Astron. Soc. Can.*, **98**, 149–153.

Fairbairn, M. 2005. Planetary photometry: The Lommel–Seeliger law. *J. Roy. Astron. Soc. Can.*, **99**, 92–93.

Falco, C.M. and Weintz Allen, A.L. 2008. *Ibn al-Haytham's contributions to optics, art, and visual literacy*. Painted Optics Symposium: Florence, Italy.

Farnham, T.L., Schleicher, D.G., and A'Hearn, M.F. 2000. The HB narrowband comet filters: Standard stars and calibrations. *Icarus*, **147**, 180–204.

Ferrin, I. 2010. Atlas of secular light curves of comets. *Planet. Space Sci.*, **58**, 365–391.

Fowles, G.R. 1975. *Introduction to Modern Optics*. 2nd Ed. Dover Publications, Inc.: New York, NY.

French, A.P. 1990. A role of history in teaching physics. In J. Roche, ed., *Physicists Look Back: Studies in the History of Physics*. IOP Publishing: Bristol, England. 111–125

Fresnel, A. 1819. Memoir on the diffraction of light. In H. Crew, ed., *The Wave Theory of Light – Memoirs by Huygens, Young and Fresnel*. American Book Company: New York, NY. 79–144.

Gold, T. 1955. The lunar surface. *Mon. Not. R. Astr. Soc.*, **115**, 585–604.

Gaffey, M.J. 1997. Surface lithologic heterogeneity of Asteroid 4 Vesta. *Icarus*, **127**, 130–157.

Geake, J.E., Geake, M., and Zellner, B.H. 1984. Experiments to test theoretical models of the polarization of light by rough surfaces. *Mon. Not. R. Astr. Soc.*, **210**, 89–112.

Geake, J.E. and Dollfus, A. 1986. Planetary surface texture and albedo from parameters plots of optical polarization data. *Mon. Not. R. Astr. Soc.*, **218**, 75–91.

Gehrels, T. 1956. Photometric studies of asteroids. V. The light-curve and phase function of 20 Massalia. *Astrophys. J.*, **123**, 331–338.

Goguen, J. 1981. *A Theoretical and Experimental Investigation of the Photometric Functions of Particulate Surfaces*. Cornell University, Ph.D. Thesis.

Gray, D.F. 1992. The inferred color index of the Sun. *Publ. Astron. Soc. Pac.*, **104**, 1035–1038.

Guinness, E. 1981. Spectral properties (0.40 to 0.75 microns) of soils exposed at the Viking 1 landing site. *J. Geophys. Res. (Solid Earth)*, **86**, 7983–7992.

Gunderson, K., Thomas, N., and Whitby, J.A. 2006. First measurements with the Physikalisches Institut Radiometric Experiment (PHIRE). *Planet. Space Sci.*, **54**, 1046–1056.

Gunn, J.E., Carr, M., Rockosi, C., et al. 1998. The Sloan Digital Sky Survey photometric camera. *Astron. J.*, **116**, 3040–3081.

Hall, J.C. (ed.) 1997. *Solar Analogs: Characteristics and Optimum Candidates*. Ed. Second Annual Lowell Observatory Fall Workshop, October 1997. Lowell Observatory.

Hapke, B. 1962. *Second Preliminary Report on Experiments Relating to the Lunar Surface*. Center Radiophys. Space Res. Rept. 127. Cornell University, Ithaca, NY.

Hapke, B. 1965. Effects of simulated solar wind on the photometric properties of rocks and powders. *Ann. N. Y. Acad. Sci.*, **123**, 711–721.

Hapke, B. 1981. Bidirectional reflectance spectroscopy, 1: Theory. *J. Geophys. Res.*, **86**, 3039–3054.

Hapke, B. 1984. Bidirectional reflectance spectroscopy, 3: Correction for macroscopic roughness. *Icarus*, **59**, 41–59.

Hapke, B. 1986. Bidirectional reflectance spectroscopy, 4: The extinction coefficient and opposition effect. *Icarus*, **67**, 264–280.

Hapke, B. 1990. Coherent backscatter and the radar characteristics of outer planet satellites. *Icarus*, **88**, 407–417.

Hapke, B. 1993. *Theory of Reflectance and Emittance Spectroscopy*. Cambridge University Press: Cambridge, UK.

Hapke, B. 1999. Scattering and diffraction of light by particles in planetary regoliths. *J. Quant. Spectrosc. Radiat. Transf.*, **61**, 565–581.

Hapke, B. 2001. Space weathering from Mercury to the asteroid belt. *J. Geophys. Res.*, **106**, 10039–10073.

Hapke, B. 2002. Bidirectional reflectance spectroscopy, 5: The coherent backscatter opposition effect and anisotropic scattering. *Icarus*, **157**, 523–534.

Hapke, B. 2008. Bidirectional reflectance spectroscopy, VI: Effects of porosity. *Icarus* **195**, 918–926.

Hapke, B. 2012. *Theory of Reflectance and Emittance Spectroscopy*. 2nd Ed. Cambridge University Press: Cambridge, UK.

Hapke, B. and Van Horn, H. 1963. Photometric studies of complex surfaces, with applications to the Moon. *J. Geophys. Res.*, **68**, 4545–4570.

Hapke, B. and Wells, E. 1981. Bidirectional reflectance spectroscopy, 2: Experiments and observations. *J. Geophys. Res.*, **86**, 3055–3060.

Hapke, B. and Blewett, D. 1991. Coherent backscatter model for the unusual radar reflectivity of icy satellites. *Nature*, **352**, 46–47.

Hapke, B., Nelson, R., and Smythe, W. 1993. The opposition effect of the Moon: The contribution of coherent backscatter. *Science*, **260**, 509–511.

Hapke, B., Nelson, R., and Smythe, W. 1998. The opposition effect of the Moon: Coherent backscatter and shadow hiding. *Icarus*, **133**, 89–97.

Hardorp, J. 1978. The Sun among the stars. *Astron. Astrophys.*, **63**, 383–390.

Hardorp, J. 1980. The Sun among the stars, III: Energy distributions of 16 northern G-type stars and the solar flux calibration. *Astron. Astrophys.*, **91**, 221–232.

Harris, D.L. 1961. Photometry and colorimetry of planets and satellites. In G.P. Kuiper and B.M. Middlehurst, eds., *Planets and Satellites*. University of Chicago Press: Chicago, IL. 272–342.

Harris, A.W., Yound, J.W., Contreiras, L., et al. 1989. Phase relations of high albedo asteroids: The unusual opposition brightening of 44 Nysa and 64 Angelina. *Icarus*, **81**, 365–374.

Hawkins, S.E., Boldt, J.D., Darlington, E.H., et al. 2007. The Mercury dual imaging system on the MESSENGER spacecraft. *Space Sci. Rev.*, **131**, 247–338.

Hawkins, S.E., Murchie, S.L., Becker, K.J., et al. 2009. In-Flight performance of MESSENGER's Mercury dual imaging system. *Proc. of SPIE Vol.* **7441**, 1–12.

Hearnshaw, J.B. 1996. *The Measurement of Starlight: Two Centuries of Astronomical Photometry*. Cambridge University Press: Cambridge, UK.

Heiken, G.H., Vaniman, D.T., and French, B.M. 1991. *The Lunar Sourcebook*. Cambridge University Press: Cambridge, UK.

Helfenstein, P. 1988. The geological interpretation of photometric surface roughness. *Icarus*, **73**, 462–481.

Helfenstein, P. and Veverka, J. 1989. Physical characterization of asteroid surfaces from photometric analysis., In R.P. Binzel, T. Gehrels, and M.S. Matthews, eds., *Asteroids II*. University of Arizona Press: Tucson, AZ. 557–593.

Henden, A.A. and Kaitchuck, R.H. 1990. *Astronomical Photometry*. 2nd Ed. Willman-Bell: Richmond, VA.

Hillier, J.K., Buratti, B.J., and Hill, K. 1999. Multispectral photometry of the Moon and absolute calibration of the Clementine UV/Vis Camera. *Icarus*, **141**, 205–225.

Hilton, J. 2005. Improving the visual magnitudes of the planets in the Astronomical Almanac, I: Mercury and Venus. *Astron. J.*, **129**, 2901–2906.

Howell, S.B. 2006. *Handbook of CCD Astronomy*. 2nd Ed. Cambridge University Press: Cambridge, UK.

van de Hulst, H.C. 1981. *Light Scattering by Small Particles*. Dover Publications, Inc.: New York, NY.

Huygens, C. 1900 [1690]. Treatise on light. In H. Crew, ed., *The Wave Theory of Light – Memoirs by Huygens, Young and Fresnel*. American Book Company: New York, NY. 1–41.

Ilardi, V. 2007. *Renaissance Vision from Spectacles to Telescopes*. American Philosophical Society: Philadelphia, PA.

International Astronomical Union (IAU). 1928. *Third General Assembly*. www.iau.org/static/resolutions/IAU1928_French.pdf.

International Astronomical Union (IAU). 1992. *Proceedings of the Twenty-First General Assembly, Buenos Aires 1991*, ed. J. Bergeron, Transactions of the IAU, Vol. XXI-B. Kluwer: Dordrecht, The Netherlands. 30–78.

International Astronomical Union (IAU). 2001. *Proceedings of the Twenty-Fourth General Assembly, Manchester 2000*, ed. H. Rickman, Transactions of the IAU, Vol. XXIV-B. ASP: San Francisco, CA. 34–59.

International Astronomical Union (IAU). 2008. *Proceedings of the Twenty-Sixth General Assembly, Prague 2006*, ed. K.A. van der Hucht, Transactions of the IAU, Vol. XXVI-B. Cambridge University Press: Cambridge, UK. 34–47.

International Astronomical Union (IAU). 2012. Resolution B2 on the re-definition of the astronomical unit of length. 28th General Assembly of the IAU. Cambridge: Cambridge University Press.

International Astronomical Union (IAU). 2015. Resolution B2 on recommended zero points for the absolute and apparent bolometric magnitude scales. 29th General Assembly of the IAU. Cambridge: Cambridge University Press.

Ivezic, Z., Tabachnik, S., Rafikov, R., et al. 2001. Solar system objects observed in the SDSS commissioning data. *Astron. J.*, **122**, 2749–2784.

Jackson, R.D., Clarke, T.R., and Moran, M.S. 1992. Bidirectional calibration results for 11 spectralon and 16 BaSO$_4$ reference reflectance panels. *Remote Sens. Environ.*, **40**, 231–239.

Johnson, H.L. 1965. The absolute calibration of the Arizona photometry. Comm. Lunar and Planetary Lab. **53**, 73–77.

Jones, J.H., Christian, C.A., and Waddell, P. 1988. Resolved CCD photometry of Pluto and Charon. *Publi. Astron. Soc. Pac.*, **100**, 489–495.

Johnson, H.L. and Morgan, W.W. 1953. Fundamental stellar photometry for standards of spectral type on the revised system of the Yerkes spectral atlas. *Astrophys. J.*, **117**, 313.

Kaasalainen, S. 2003. Laboratory photometry of planetary regolith analogs, I: Effects of grain and packing properties on the opposition effect. *Astron. Astrophys.*, **409**, 765–769.

Kaasalainen, M. and Torppa, J. 2001. Optimization methods for asteroid lightcurve inversion. *Icarus*, **153**, 24–36.

Kaasalainen, M., Lamberg, L., Lumme, K., and Bowell, E. 1992a. Interpretation of lightcurves of atmosphereless bodies, I: General theory and new inversion schemes. *Astron. Astrophys.*, **259**, 318–332.

Kaasalainen, M., Lamberg, L., Lumme, K., and Bowell, E. 1992b. Interpretation of lightcurves of atmosphereless bodies, II: Practical aspects of inversion. *Astron. Astrophys.*, **259**, 333–340.

Kaasalainen, M., Torppa, J., and Piironen, J. 2002. Models of twenty asteroids from photometric data. *Icarus*, **159**, 369–395.

Kaler, J.B. 1997. *Stars and Their Spectra: An Introduction to the Spectral Sequence*. Cambridge University Press: Cambridge, UK.

Kaplan, G.H. 2005. *The IAU Resolutions on Astronomical Reference Systems, Time Scales, and Earth Rotation Models*. US Naval Observatory Circular No. 179. USNO: Washington, D.C.

Karttunen, H. 1989. Modelling asteroid brightness variations, I: Numerical methods. *Astron. Astrophys.*, **208**, 314–319.

Kaydash, V.G., Gerasimenko, S., Shkuratov, Yu.G., et al. 2010. The phase ratios of the color index: Mapping of two regions of the near side of the Moon. *Sol. Syst. Res.*, **44**, 267–280 [Translated from Astronomicheskii Vestnik, 2010, **44**, 291–304].

Keck Telescope and Facility Instrument Guide. 2002. www2.keck.hawaii.edu/observing/kecktelgde/ktelinstupdate.pdf.

Keller, L.P. and McKay, D.S. 1993. Discovery of vapor deposits in the lunar regolith. *Science*, **261**, 1305–1307.

Keller, M.R., Ernst, C.M., Denevi, B.W., et al. 2013. Time-dependent calibration of MESSENGER's wide-angle camera following a contamination event. *Lunar Planet. Sci.*, **44**, Abstract 2489.

King, H.C. 1955. *The History of the Telescope*. Charles Griffin & Co. Ltd: London.

Kitchin, C.R. 2003. *Astrophysical Techniques*. 4th Ed. Taylor & Francis: Boca Raton, FL.

Kolokolova, L., Hanner, M.S., Levasseur-Regourd, A., and Gustafson, B. 2004. Physical properties of cometary dust from light scattering and thermal emission. In M.C. Festou, H.U. Keller, and H.A. Weaver, eds., *Comets II*. University Arizona Press: Tucson, AZ. 577–604.

Kortum, G. 1969. *Reflectance Spectroscopy: Principles, Methods, Applications*. (J.E. Lohr translation from German). Springer-Verlag: Berlin, Germany.

Kreslavsky, M.A. and Shkuratov, Y.G. 2003. Photometric anomalies of the lunar surface: Results from Clementine data. *J. Geophys. Res.*, **108**, 5015.

Kreslavsky, M.A., Shkuratov, Y.G., Velikodsky, Yu.I., et al. 2000. Photometric properties of the lunar surface derived from Clementine observations. *J. Geophys. Res.*, **105** (E8), 20,281–20,296.

Kreslavsky, M.A., Shkuratov, Y.G., Kaydash, V.G., et al. 2001. Lunar opposition spike observed by Clementine NIR Camera: Preliminary results. *Lunar Planet. Sci.*, 32-rd. LPI: Houston. Abstract #1140. 37.

Kreyszig, E. 1983. *Advanced Engineering Mathematics*. 5th Ed. John Wiley & Sons: New York, NY.

Lambert, J.H. 1760. *Photometria, sive de Mensura et Gradibus Luminis, Colorum et Umbrae*. Augsburg. Germany: Klett.

Landolt, A.U. 1983. UBVRI photometric standard stars around the celestial equator. *Astron. J.*, **88**, 439–460.

Landolt, A.U. 1992. UBVRI photometric standard stars in the magnitude range $11.5 < V < 16.0$ around the celestial equator. *Astron. J.*, **104**, 340–491.

Landolt, A.U. and Uomoto, A.K. 2007a. Optical multicolor photometry of spectrophotometric standard stars. *Astron. J.*, **133**, 768–790.

Landolt, A.U. and Uomoto, A.K. 2007b. Erratum: "Optical multicolor photometry of spectrophotometric standard stars." *Astron. J.*, **133**, 2429.

Lane, A.P. and Irvine, W.M. 1973. Monochromatic phase curves and albedos for the lunar disk. *Astron. J.*, **78**, 267–277.

Lester, T.P., McCall, M.L., and Tatum, J.B. 1979. Theory of planetary photometry. *J. Roy. Astron. Soc. Can.*, **73**, 233–257.

Li, J.-Y., Reddy, V., Nathues, A., et al. 2016. Surface albedo and spectral variability of Ceres. *Astrophys. J. Lett.*, **817**, 2, L22.

Lindberg, D.C. 1981. *Theories of Vision from Al-kindi to Kepler*. University of Chicago Press: Chicago, IL.

Lumme, K. and Bowell, E. 1981. Radiative transfer in the surfaces of atmosphereless bodies, I: Theory, *Astron. J.*, **86**, 1694–1704.

Mach, E. 1926. *The Principles of Physical Optics. An Historical and Philosophical Treatment*. Translated by J.S. Anderson and A.F.A. Young. Dover: New York, NY.

Malet, A. 2010. Kepler's legacy: Telescopes and geometrical optics, 1611–1669. In A. Van Helden, S. Dupre, R. van Gent, and H. Zuidervaart, eds., *Origins of the Telescope*. Royal Netherlands Academy of Arts and Sciences: Amsterdam. 281–300.

Mallama, A. 2007. The magnitude and albedo of Mars. *Icarus*, **192**, 404–416.

Mallama, A., Wang, D., and Howard, R.A. 2002. Photometry of Mercury from SOHO/LASCO and Earth. *Icarus*, **155**, 253–264.

Mallama, A., Wang, D., and Howard, R.A. 2006. Venus phase function and forward scattering from H_2SO_4. *Icarus*, **182**, 10–22.

Markov, A.V. 1924. Les particularites dans la refiexion de la lumiere par la surface de la lune. *Astron. J. Nachr.*, **221**, 65–78.

Markowitz, W., Hall, R.G., Essen, L., and Parry, J.V.L. 1958. Frequency of cesium in terms of Ephemeris Time. *Phys. Rev. Lett.*, **1**, 105–107.

Marov, M.Y. 2005. *Mikhail Lomonosov and the discovery of the atmosphere of Venus during the 1761 transit*. Transits of Venus: New Views of the Solar System and Galaxy, Proceedings of IAU Colloquium #196, held June 7–11, 2004 in Preston, U.K. Edited by D.W. Kurtz. Cambridge University Press: Cambridge, UK. 209–219.

Marsden, B. 1985. Notes from the IAU General Assembly. Minor Planet Circular 10193. International Astronomical Union Minor Planet Center.

Marsden, B. and Roemer, E. 1983. Basic information and references. In L.L. Wilkening, ed., *Comets*. University of Arizona Press: Tucson, AZ. 707–736.

Masiero, J.R., Mainzer, A.K., Grav, T., et al. 2012. A revised asteroid polarization-albedo relationship using WISE/NEOWISE data. *Astrophys. J.*, **749**, 6. doi:101088/0004-637X/749/2/104.

Masin, S.C., Zudini, V., and Antonelli, M. 2009. Early alternative derivations of Fechner's Law. *J. Hist. Behav. Sci.*, **45**, 56–65.

McCord, T.B. 1969. Color differences on the lunar surface. *J. Geophys. Res.*, **74**, 3131–3142.

McEwen, A.S. 1996. A precise lunar photometric function. *LPSC*, **27**, 841 (abstract).

McEwen, A.S., Eliason, E.M., Bergstrom, J.W., et al. 2007. Mars reconnaissance orbiter's high resolution imaging science experiment (HiRISE). *J. Geophys. Res.*, **112**, E05S02. doi:10.1029/2005JE002605.

McGuire, A. and Hapke, B. 1995. An experimental study of light scattering by large, irregular particles. *Icarus*, **113**, 134–155.

Meeus, J. 2000. *Astronomical Algorithms*. Wilman-Bell: Richmond, VA.

Meisel, D.D. and Morris, C.S. 1983. Comet head photometry: Past, present, and future. In L. L. Wilkening, ed., *Comets*. University of Arizona Press: Tucson, AZ. 413–432.

Mersenne, M. 1651. *L'optique et la catoptrique*. Paris: Langlois.

Minnaert, M. 1941. The reciprocity principle in lunar photometry. *Astrophys. J.*, **93**, 403–410.

Minnaert, M. 1961. Photometry of the moon in planets and satellites. ed. G.P. Kuiper and B.M. Middlehurst. University of Chicago Press: Chicago, IL. 213–248.

Mishchenko, M.I. 1992. The angular width of the coherent back-scatter opposition effect – an application to icy outer planet satellites. *Astrophys. Space Sci.*, **194**, 327–333.

Mishchenko, M.I. 2013. 125 years of radiative transfer: Enduring triumphs and persisting misconceptions. *AIP Conf. Proc.*, **1531**, 11–18.

Mishchenko, M.I. and Dlugach, J.M. 1993. Coherent backscatter and the opposition effect for E-type asteroids. *Planet. Space Sci.*, **41**, 173–181.

Moore, P. 1983. *The Guinness Book of Astronomy Facts and Feats*. Guinness Superlatives Ltd.: Enfield, Middlesex.

Mueller, B., Samarasinha, N.H., and Belton, M.J. 2002. The diagnosis of complex rotation in the lightcurve of 4179 Toutatis and potential applications to other asteroids and bare cometary nuclei. *Icarus*, **158**, 305–311.

Muinonen, K. 1990. Scattering of light by solar system dust: The coherent backscatter phenomenon. Proceedings of the Finnish Astronomical Society, pp. 12–15

Muinonen, K., Penttil, A., Cellino, A., et al. 2009. Asteroid photometric and polarimetric phase curves: Joint linear-exponential modeling. *Meteor. Planet. Sci.*, **44**, 1937–1946.

Muinonen, K., Belskaya I.N., Cellino, A., et al. 2010. A three-parameter magnitude phase function for asteroids. *Icarus*, **209**, 542–555.

Muller, G. 1893. Ueber die Lichtstarke des Planeten Merkur. *Astron. Nachr.*, **133**, 47.

Mustard, J.F. and Pieters, C.M. 1987. Quantitative abundance estimates from bidirectional reflectance measurements. *J. Geophys. Res.*, **92**, E617–E626.

Mustard, J.F. and Pieters, C.M. 1989. Photometric phase functions of common geologic materials and applications to quantitative analysis of mineral mixtures. *J. Geophys. Res.*, **94**, 13,619–13,634.

Naranen, J., Kaasalainen, S., Peltoniemi, J., et al. 2004. Laboratory photometry of planetary regolith analogs, II: Surface roughness and extremes of packing density. *Astron. Astrophys.*, **426**, 1103–1109.

Nella, J., Atcheson, P.D., Atkinson, B., et al. 2004. James Webb Space Telescope (JWST) Observatory architecture and performance in, Optical, Infrared, and Millimeter Space Telescopes. Edited by J.C. Mather. *Proc. SPIE*, **5487**, 576–587.

Nelson, R., Hapke, B., Smythe, W., and Horn, L. 1998. Phase curves of selected particulate materials: The contribution of coherent backscattering to the opposition surge. *Icarus*, **131**, 223–230.

Nelson, R., Hapke, B., Smythe, W., and Spilker, L. 2000. The opposition effect in simulated planetary regolith. Reflectance and circular polarization ratio change at small phase angle. *Icarus*, **147**, 545–558.

Newton, I. 1704. *Opticks*. Dover Publications: New York, NY.
Illuminating Engineering Society (IES) [Committee on Nomenclature]. 2010. *Nomenclature and Definitions for Illuminating Engineering*. RP-16-10. IES: New York, NY.
Oke, J.B., and Gunn, J.E. 1983. Secondary standard stars for absolute spectrophotometry. *Astrophys. J.*, **266**, 713–717.
Olivier, S.S., Seppala L., Gilmore, K., and the LSST camera team. 2008. Optical Design of the LSST Camera, LLNL-CONF-405460, Advanced Optical and Mechanical Technologies in Telescopes and Instrumentation. Edited by E. Atad-Ettedgui and D. Lemke. *Proc. SPIE*, **7018**, Article id. 70182G, 9 pp.
Onehag et al. 2011. M67-1194, an unusually Sun-like solar twin in M67. *Astronomy Astrophys.*, **528**, id.A85, 11pp.
Opik, E. 1924. Photometric measures on the moon and the earth-shine. *Publ. Astron. Obs. Tartu*, **26**, 1–68.
Osborn, W.H., A'Hearn, M.E., Carsenty, U., et al. 1990. Standard stars for photometry of comets. *Icarus*, **88**, 228–245.
Ostro, S.J., Connelly, R., and Dorogi, M. 1988. Convex-profile inversion of asteroid lightcurves: Theory and applications. *Icarus*, **75**, 30–63.
Oszkiewicz, D.A., Bowell, E., Wasserman, L.H., et al. 2012. Asteroid taxonomic signatures from photometric phase curves. *Icarus*, **219**, 283–296.
Pieters, C.M. 1999. The Moon as a spectral calibration standard enabled by lunar samples: The Clementine example. Workshop on New Views of the Moon II, Flagstaff, AZ, abstract 8025.
Pinet, P.C., Cord, A., Daydou, F., et al. 2001. Influence of linear versus non-linear mixture on bidirectional reflectance spectra using a laboratory wide field spectral imaging facility. *Lunar Planet. Sci.*, **XXXII**, abstract 1552.
Pommerol, A., Thomas, N., Portyankina, G., and Jost, B. 2011. Photometry of icy planetary analogs: First results of the PHIRE-2 experiment. EPSC-DPS Joint Meeting 2011, held October 2–7, 2011 in Nantes, France. 463.
Potsdam photometric observations. 1894. *The Observatory*, **17**, 240–266.
Rashed, R. 1990. A pioneer in anaclastics: Ibn Sahl on burning mirrors and lenses. *Isis*, **81**, 464–491.
Rayner, J.T., Toomey, D.W., Onaka, P.M., et al. 1998. SpeX: A medium-resolution IR spectrograph for IRTF. *Proc. SPIE*, **3354**, 468–479.
Reddy, V., Li, J.-Y., Le Corre, L., et al. 2013. Comparing Dawn, Hubble Space Telescope, and ground-based interpretations of (4) Vesta. *Icarus*, **226**, 1103–1114.
Robinson, M.S., Brylow, S.M., Tschimmel, M., et al. 2010. Lunar reconnaissance orbiter camera (LROC) instrument overview. *Space Sci. Rev.*, **150**, 81–124.
Robinson, M.S., Thomas, P.C., Veverka, J., Murchie, S., and Carchich, B. 2001. The nature of ponded deposits on Eros. *Nature*, **413**, 396–400.
Roemer, O. 1677. On the motion of light. *Phil. Trans. Roy. Soc.*, **12**, 397–398.
Rosenbush, V., Kiselev, N., Avramchuk, V., and Mishenko, M. 2002. Photometric and polarimetric opposition phenomenon exhibited by solar system bodies. In G. Videen and M. Kocifaj, eds., *Optics of Cosmic Dust*. Kluwer Academic Publishers: Dordrecht, The Netherlands. 191–224.
Rosenbush, V., Shevchenko, V.G., Kiselev, N.N., et al. 2009. Polarization and brightness opposition effects for the E-type asteroid 44 Nysa. *Icarus*, **201**, 655–665.

Rossotti, H. 1983. *Colour: Why the World Isn't Grey*. Princeton University Press: Princeton, NJ.

Russell, H.N. 1906. On the light-variations of asteroids and satellites. *Astrophys. J.*, **24**, 1–18.

Russell, H.N. 1916a. The stellar magnitudes of the Sun, Moon, and Planets. *Astrophys. J.*, **43**, 103–129.

Russell, H.N. 1916b. On the albedo of the planets and their satellites. *Astrophy. J.*, **43**, 173–196.

Sasaki, S., Kurahashi, E., Yamanaka, C., and Nakamura, K. 2003. Laboratory simulation of space weathering: Changes of optical properties and TEM/ESR confirmation of nanophase metallic iron. *Adv. Space Res.*, **31**, 2537–2542.

Sato, H., Robinson, M.S., Hapke, B., Denevi, B.W., and Boyd, A.K. 2014. Resolve Hapke parameter maps of the Moon. *J. Geophys. Res. Planets*, **119**, 1775–1805.

Schleicher, D.G. and Farnham, T.L. 2004. Photometry and imaging of the coma with narrowband filters. In M.C. Festou, H.U. Keller, and H.A. Weaver, eds., *Comets II*. University of Arizona Press: Tucson, AZ. 449–470.

Schleicher, D.G., Bair, A.N., Sackey, S., et al. 2015. The evolving photometric light-curve of comet 1P/Halley's coma during the 1985/86 apparitions. *Astron. J.*, **150**.

Schroder, S.E., Grynko, Ye., Pommerol, A., et al. 2014. Laboratory observations and simulations of phase reddening. *Icarus*, **239**, 201–216.

von Seeliger, H. 1887. Zur Theorie der Beleuchtung der grossen Planeten insbesondere des Saturn. *Abh. Bayer. Akad. Wiss. Math. Naturwiss. Kl.*, **16**, 405–516.

von Seeliger, H. 1895. Theorie der Beleuchtung staubformiger kosmischen Masses insbesondere des Saturinges. *Abhandl. Bayer. Akad. Wiss. Math. Naturw. Kl. II*, **18**, 1–72.

Shaw, A., Daly, M.D., Cloutis, E., et al. 2016. Reflectance properties of grey-scale Spectralon® as a function of viewing angle, wavelength, and polarization. *Int. J. Rem. Sens.*, **37**, 2510–2523.

Shepard, M.K. 2001. *The Bloomsburg University Goniometer (B.U.G.) Laboratory: An Integrated Laboratory for Measuring Bidirectional Reflectance Functions*. 32nd Annual Lunar and Planetary Science Conference, March 12–16, 2001, Houston, TX, abstract 1015.

Shepard, M.K. 2015. *Asteroids: Relics of Ancient Time*. Cambridge University Press: Cambridge, UK.

Shepard, M.K. and Campbell, B.A. 1998. Shadows on a planetary surface and implications for photometric roughness. *Icarus*, **134**, 279–291.

Shepard, M.K. and Helfenstein, P. 2007. A test of the Hapke photometric model. *J. Geophys. Res.*, **112**, E03001. 17pp.

Shepard, M.K., Campbell, B.A., Bulmer, M.H., et al. 2001. The roughness of natural terrain: A planetary and remote sensing perspective. *J. Geophys. Res.*, **106**, 32,777–32,795.

Shevchenko, V.G. 1997. Analysis of asteroid brightness phase relations. *Sol. Syst. Res.*, **31**, 219–224.

Shevchenko, V.G., Chiorny, V.G., Gaftonyuk, N.M., et al. 2008. Asteroid observations at low phase angles, III: Brightness behavior of dark asteroids. *Icarus*, **196**, 601–611.

Shkuratov, Y. 1989. A new mechanism for the negative polarization of light scattered by the solid surfaces of cosmic bodies. *Astron. Vestnik*, **23**, 176–180 (in Russian).

Shkuratov, Y., Kreslavsky, M.A., Ovcharenko, A.A., et al. 1999. Opposition effect from Clementine data and mechanisms of backscatter. *Icarus*, **141**, 132–155.

Shkuratov, Y.G., Ovcharenko, A., Zubko, E., et al. 2002. The opposition effect and negative polarization of structural analogs for planetary regoliths. *Icarus*, **159**, 396–416.

Shkuratov, Y., Kaydash, V., Gerasimenko, S., et al. 2010. Probable swirls detected as photometric anomalies in Oceanus Procellarum. *Icarus*, **208**, 20–30.

Shkuratov, Y., Kaydash, V., Korokhin, V., et al. 2011. Optical measurements of the Moon as a tool to study its surface. *Planet. Space Sci.*, **59**, 1326–1371.

Shorthill, R.W. and Saari, J.M. 1965. Non-uniform cooling of the eclipsed moon: A listing of thirty prominent anomalies. *Science*, **150**, 210–212.

Shorthill, R.W., Saari, J.M., Baird, F.E., and LeCompte, J.R. 1969. *Photometric Properties of Selected Lunar Features*. NASA Contractor Report CR-1429. 405 pp.

Sidgwick, J.B. 1971. *The Amateur Astronomer's Handbook*. Dover Publications: New York, NY.

Sierks, H., Keller, H.U., Jaumann, H., et al. 2012. The Dawn framing camera. In C. Russell and C. Raymond, eds., *The Dawn Mission to Minor Planets 4 Vesta and 1 Ceres*. Springer: New York, NY.

Simons, H. 1964. Is the Moon being skinned? *New Scientist*, **413**, 163.

Skrutskie, M.F., Cutri, R.M., Stiening, R., et al. 2006. The two micron all sky survey (2MASS). *Astron. J.*, **131**, 1163–1183.

Smith, J.A., Tucker, D.L., Stephen, K., et al. 2002. The u'g'r'i'z' standard star system. *Astron. J.*, **123**, 2121–2144.

Soderblom, D.R. and King, J.R. 1998. Solar-type stars: Basic information on their classification and characterization. In J.C. Hall, ed., *Solar Analogs: Characteristics and Optimum Candidates*. Second Annual Lowell Observatory Fall Workshop, October 1997. Lowell Observatory: Flagstaff, AZ. 41–60.

Springsteen, A.W. 1989. A novel class of Lambertian reflectance materials for remote sensing application. *Opt. Radiat. Meas. II*, **1109**: 133–141.

Springsteen, A.W. 1999. Standards for the measurement of diffuse reflectance - an overview of available materials and measurement laboratories. *Anal. Chim. Acta*, **380**, 379–390.

Standish, E.M. 1998. Time scales in the JPL and CfA ephemerides. *Astron. Astrophys.*, **336**, 381–384.

Staubermann, K. 2000. The trouble with the instrument: Zöllner's photometer. *J. Hist. Astron.*, **31**, 323–338.

Sykes, M.V., Cutri, R.M., Fowler, J.W., et al. 2000. The 2MASS asteroid and comet survey. *Icarus*, **146**, 161–175.

Sykes, M.V., Cutri, R.M., Fowler, J.W., et al. 2010. 2MASS asteroid and comet survey V2.0. NASA Planetary Data System, EAR-A-I0054/I0055-5-2MASS-V2.0.

Taylor, R.C., Gehrels, T., and Silvester, A.B. 1971. Minor planets and related objects. VI. Asteroid (110) Lydia. *Astron. J.*, **76**, 141–146.

Thompson, B. 1794. A method of measuring the comparative intensities of the light emitted by luminous bodies. *Philos. Trans.*, **84**, 67. Rumfords Collected Works (Boston, 1875, vol. 4, p. 1).

Thompson, T.W. and Dyce, R.B. 1966. Mapping of Lunar reflectivity at 70 centimeters. *J. Geophys. Res.*, **71**, 4843–4853.

Thorpe, T.E. 1976. The Viking Orbiter cameras' potential for photometric measurement. *Icarus*, **27**, 229–239.

Thuan, T.X. and Gunn, J.E. 1976. The new four-color intermediate band photometric system. *Publ. Astron. Soc. Pac.*, **88**, 543–547.

United States Naval Observatory (USNO). http://aa.usno.navy.mil/faq/docs/ICRS_doc.php.

van Diggelen, J. 1959. Photometric properties of lunar crater floors. *Rech. Obs. Utrecht*, **14**, 2.

Veverka, J. 1971a. The meaning of Russell's Law. *Icarus*, **14**, 284–285.

Veverka, J. 1971b. The Physical Meaning of Phase Coefficients. In T. Gehrels, ed., *Physical Studies of Minor Planets*. Proceedings of IAU Colloq. 12, held in Tucson, AZ, National Aeronautics and Space Administration, SP 267, Washington, DC. 79–90.

Vsekhsvyatskii, S.K. 1964. Physical characteristics of comets. *Soviet Astron.-AJ*, **8**, 429–431; also NASA TT F-80, OTS 62-11031.

Walsh, J.W.T. 1926. *Photometry*. Constable and Company: London, UK.

Wade, R.A., Hoessel, J.G., Elias, J.H., et al. 1979. A two-color photometric system for the near-infrared. *Publ. Astron. Soc. Pac.*, **91**, 35–40.

Warner, B.D. 2006. *A Practical Guide to Lightcurve Photometry and Analysis*. Springer: New York, NY.

Warell, J. 2004. Properties of the Hermean regolith, IV: Photometric parameters of Mercury and the Moon contrasted with Hapke modelling. *Icarus*, **167**, 271–286.

Warell, J. and Bergfors, C. 2008. Mercury's integral phase curve: Phase reddening and wavelength dependence of photometric quantities. *Planet. Space Sci.*, **56**, 1939–1948.

Weber, E.H. 1834. *De pulsu, resorption, auditu et tactu*. Annotationes anatomicae et physiologicae. Leipzig: Koehler.

Wehner, G.K., Rosenberg, D., and Kenknight, C.E. 1963. Modification of the lunar surface by the solar-wind bombardment. *Planet. Space Sci.*, **11**, 1257–1261.

Wilson, R.C. and Hudson, H.S. 1991. The Sun's luminosity over a complete solar cycle. *Nature*, **351**, 42–44.

Wood, C.A. 1987. *Rotation Period of Halley's and Other Comets*. Abstract 1560. LPSC XVIII, Houston, TX.

Wyckoff, S. 1983. Overview of comet observations. In L.L. Wilkening, ed., *Comets*. Arizona Press: Tucson, AZ. 3–55.

Young, T. 1804. Bakerian Lecture: Experiments and calculations relative to physical optics. *Philos. Trans. Roy. Soc.*, **94**, 1–16.

Zöllner, J.C.F. 1865. *Photometrische untersuchungen mit Besonderer Rücksicht auf die Physische Beschaffenheitt der Himmelskörper*. (Photometric investigations with special reference to the physical nature of celestial bodies) (German). Leipzig: Verlag von Wilhelm Engelmann.

Index

absolute magnitude, 51–2, 171
air mass, 91
Airy pattern, 72
Airy, G., 72
Akimov disk function, 150
Alhazen, 3
alt-azimuth mount, 65
Ambartsumian, V., 160
apparent magnitude, 50
Arago, F., 4
Area law, 22, 158, 175, 192
asymmetry parameter, 108, 110

Bartholin, R., 5
Barycentric Celestial Reference System, 31
Barycentric Dynamical Time, 41
Beer, A., 13
Beer's law, 13, 115
bidirectional reflectance, 135
bidirectional reflectance distribution function, 26, 136
bi-hemispherical reflectance, 119
birefringent, 5, 100
blackbody, 5
bolometric magnitude, 62
Bond albedo, 119, 154
Bond, G., 154
Bouguer, P., 11
Bowell, E., 164

candela, 18
Cassegrain, L., 68
Celsius, A., 9
Ceraski, V., 21
Chandrasekhar, S., 160

charge-coupled device
 bias, 82
 blooming, 78
 construction, 77–8
 dynamic range, 81, 83
 fill factor, 78
 frame-transfer, 84
 full-frame, 84
 gain, 81
 interline, 84
 matching with telescope, 87
 noise, 82
 pixel, 77
 quantum efficiency, 78
 readout, 79
 well capacity, 78
Chrétien, H., 70
chromatic aberration, 65
Clementine
 image calibration, 226
 orbit, 226
 problems with calibration, 227
 UVVIS camera, 224
coherent backscattering opposition effect, 125, 215
color index, 56
comatic aberration, 68
comet continuum, 173
comet emission spectrum, 173
comet magnitude, 172
comparison photometer, 15
conjunction, 34
cross-section for scattering, 116

Dawes, W., 73
Dawes' limit, 73

day
  Julian, 41
  sidereal, 42
  solar, 39
  standard, 40
diffraction, 4, 68
directional-hemispherical reflectance, 138
disk function, 140

ecliptic coordinate system, 34
effective diameter, 175
elongation, 34
equatorial coordinate system, 33
equatorial mount, 65
Euclid, 2
excess magnitude, 188
extinction photometer, 9, 14
eye physiology, 6

Fabry, C., 21
fairy castle structure, 27
Fechner, G., 8
field-of-view, 71, 87
Flamsteed, J., 9
flux density, 46
focal length, 67
Fresnel reflection, 101
Fresnel, A., 4, 101

genetic algorithms, 169
geometric albedo, 153, 171
geometric optics, 107
Goguen, J., 164
Gold, T., 25
goniometer, 26, 112, 206, 208, 213
gradient search inversion, 168
Greenstein, J., 109
Greenwich Mean Time, 40
grid search inversion, 168

Hall, C., 65
Halley, E., 10
Hapke model
  disk resolved, 160
  integral phase function, 163
  Mercury, 202
  opposition surge, 161, 217
  surface roughness, 162
  test of, 213
Hapke, B., 25
Hardorp, J., 61
Harvard Photometry, 18
Henyey, L., 109

Henyey-Greenstein function, 110, 112
Herschel photometer, 16
Herschel, J., 22
Herschel, W., 16
H-functions, 120, 160
HG magnitude system, 189
HG_1G_2 magnitude system, 192
Hipparchus Celestial Reference Frame, 32
Huygens, C., 3

Ibn Sahl, 100
instantaneous field-of-view, 88, 209
integral phase function, 153, 155, 163, 191
interference effect, 4
International Atomic Time, 40
International candle, 18
International Celestial Reference System, 31
inverse square law, 48
irradiance, 45
isotropic multiple-scattering approximation, 120
isotropic scattering, 109, 119

Jones vectors, 97
Jones, C., 97

Kaasalainen, M., 179
Kempf, P., 19
Kepler, J., 3
Kuiper, G., 24

Lambert, J., 11
Lambert's cosine emission law, 13
Lambert's cosine incidence law, 13
Lambertian reflectance, 155, 182
Lambertian standard, 132, 205
Lambertian surface, 26, 132, 145, 153, 171
Landolt, A., 61
LASCO, 184, 200
Leavitt, H., 61
Legendre polynomials, 109
light
  absorption, 103
  extinction, 104, 114
  polarization, 96, 102
  refraction, 100
  wave notation, 95
light-corrected time, 44
lightcurve, 175
  amplitude, 175
  aspect, 175
  Fourier series, 177
  harmonic order, 177

rotation period, 176
shape from, 178
limb-darkening/brightening, 145
limiting magnitude, 74
limiting resolution, 73
local hour angle, 43
Lommel, E., 146
Lommel–Seeliger model, 117, 146, 148, 157
Lomonosov, M., 181
Lorenz, L., 106
luminance, 11
luminance coordinates, 36
luminance equator, 142
Lumme, K., 164
Lumme–Bowell model
  disk resolved, 165
  HG system derivation, 189
  opposition surge function, 165
  roughness phase function, 165
Lunar-Lambert law, 148

Maksutov, D., 70
Maurolico, F., 11
Mercury, 201, 217
Mercury Dual Imaging System, 218
meridian photometer, 18
Mersenne, M., 3
MESSENGER, 202, 217
Mie scattering, 106
Mie, G., 106
Minnaert function, 149
Müller, G., 19
multiple scattering, 118

Nasmyth, J., 69
Newton, I., 3, 66
normal albedo, 138
null instrument, 11, 15
Nyquist, H., 89

opposition, 34
opposition surge, 121, 187–8
  lunar, 22, 228
optical depth, 115
Ostro, S., 179

particle phase function, 107
Peckham, J., 3
phase angle, 35
phase coefficient, 187
phase curve, 52
  asteroids, 185
  lunar, 22

Mercury, 184, 202, 217
physical causes, 192
planetary table, 185
relationship to albedo, 194
slope parameter, 190
$V(1,0)/\beta$ system, 187
Venus, 52, 181
phase integral, 154, 191
phase reddening, 202, 228
photography, 75
Photometria, 11
photometric equator, 38
photometric function, 140
photometric laws, 12–13
photometric reduction
  absolute photometry, 89
  aperture method, 90, 201
  differential photometry, 89
  extinction, 91
  instrumental magnitude, 91
  noise removal, 91
  normalization, 91
  PSF profile fitting, 90
  transform function, 93
photometric system
  absolute/AB, 60
  comet IHW/HB, 58, 173
  definition, 55
  Gunn griz, 57
  JHK, 57
  UBVRI, 55
photomultiplier, 75
photopic vision, 7
physical albedo, 153
Pickering, E., 18–19, 21, 61
Planck Function, 5
Planck, M., 5
plate scale, 88
Pogson Interval, 10
Pogson, N., 10
point-spread function, 72, 89
Polaris, 19
polarization phase curve
  asteroids, 198
  comets, 194
  conventions, 195
  degree of polarization, 196
  laboratory studies, 196
  minimum, 196
  negative branch, 196
Prichard, C., 19
principle axis rotation, 180
Ptolemy, 2, 9

Purkinje effect, 7
Purkyne, J., 7
pushbroom scanner, 85
pushframe scanner, 85

radiance, 49
radiance coefficient, 137
radiance factor, 26, 136, 220
radiant intensity, 48
radiative transfer, 114
Rayleigh criterion, 73
Rayleigh scattering, 105, 109
reciprocity, 140, 145, 147
reduced magnitude, 51, 170, 174
resolution, 88
Riccioli, G., 22
Ritchey, G., 70
Roemer, O., 4
Rumford-Lambert photometer, 16
Russell, H., 20, 131, 155, 178
Russell's law, 155

sampling parameter, 88
scattering angle, 107
Schmidt, B., 70
scotopic vision, 7
secular light curve, 179
Seeliger, H., 121
Seidel, P., 9
shadow hiding opposition effect, 123, 215
signal-to-noise ratio, 54
single scattering albedo, 116, 215
size parameter, 105
Snell, W., 100
Snell's Law, 100
solar analog, 62
solar magnitude, 21, 172
solid angle, 47
space weathering, 28, 237
spacecraft photometry
  data compression, 210
  processing levels, 211
  resolution, 209
  space environment, 210
spherical aberration, 68
spherical albedo, 154
SPICE kernals, 211
standard stars

Landolt, 61
North Polar Sequence, 61
Polaris, 60
Stokes parameters, 99
Stokes, G., 99
Strutt, J. Lord Rayleigh, 73, 105
surface coordinates, 36
surface roughness
  fractal surfaces, 128
  shadowing, 129
  stationary surface, 127

telescope
  Cassegrain, 68
  catadioptric, 70, 225
  coudé, 69
  Maksutov-Cassegrain, 70
  Nasmyth, 69
  Newtonian, 67
  prime focus, 68
  reflecting, 66
  refractor, 65
  Ritchey-Chrétien, 70
  Schmidt-Cassegrain, 70
  three-mirror anastigmat, 70
telescopic seeing, 74
Terrestrial Time, 40
Thompson, B., 15
Titan, 103

Universal Time, 40

van Horn, H., 26
Vega, 19
vidicon, 76
vignetting, 87
visual geometric albedo, 153
visual magnitude, 2, 9

Weber, E., 8
Weber-Fechner law, 8
wedge photometer, 15

Young, T., 4

Zöllner photometer, 17–18
Zöllner, J., 17